AF705221

The Projective Heat Map

Mathematical
Surveys
and
Monographs

Volume 219

The Projective Heat Map

Richard Evan Schwartz

American Mathematical Society
Providence, Rhode Island

EDITORIAL COMMITTEE

Robert Guralnick, Benjamin Sudakov
Michael A. Singer, Chair, Constantin Teleman
Michael I. Weinstein

2010 *Mathematics Subject Classification.* Primary 26A18, 37E30, 37B05, 51M15.

For additional information and updates on this book, visit
www.ams.org/bookpages/surv-219

Library of Congress Cataloging-in-Publication Data

Names: Schwartz, Richard Evan.
Title: The projective heat map / Richard Schwartz.
Description: Providence, Rhode Island : American Mathematical Society, [2017] | Series: Mathematical surveys and monographs ; volume 219 | Includes bibliographical references.
Identifiers: LCCN 2016037737 | ISBN 9781470435141 (alk. paper)
Subjects: LCSH: Mappings (Mathematics) | Projective spaces. | Geometry, Projective. | AMS: Real functions – Functions of one variable – Iteration. msc | Dynamical systems and ergodic theory – Low-dimensional dynamical systems – Homeomorphisms and diffeomorphisms of planes and surfaces. msc | Dynamical systems and ergodic theory – Topological dynamics – Transformations and group actions with special properties (minimality, distality, proximality, etc.). msc | Geometry – Real and complex geometry – Geometric constructions. msc
Classification: LCC QA360 .S39 2017 | DDC 516/.154–dc23 LC record available at https://lccn.loc.gov/2016037737

Copying and reprinting. Individual readers of this publication, and nonprofit libraries acting for them, are permitted to make fair use of the material, such as to copy select pages for use in teaching or research. Permission is granted to quote brief passages from this publication in reviews, provided the customary acknowledgment of the source is given.

Republication, systematic copying, or multiple reproduction of any material in this publication is permitted only under license from the American Mathematical Society. Permissions to reuse portions of AMS publication content are handled by Copyright Clearance Center's RightsLink® service. For more information, please visit: `http://www.ams.org/rightslink`.

Send requests for translation rights and licensed reprints to `reprint-permission@ams.org`.

Excluded from these provisions is material for which the author holds copyright. In such cases, requests for permission to reuse or reprint material should be addressed directly to the author(s). Copyright ownership is indicated on the copyright page, or on the lower right-hand corner of the first page of each article within proceedings volumes.

© 2017 by the American Mathematical Society. All rights reserved.
The American Mathematical Society retains all rights
except those granted to the United States Government.
Printed in the United States of America.

∞ The paper used in this book is acid-free and falls within the guidelines
established to ensure permanence and durability.
Visit the AMS home page at `http://www.ams.org/`

10 9 8 7 6 5 4 3 2 1   22 21 20 19 18 17

Contents

Preface

There is a simple and well-known construction which starts with one polygon and returns a new polygon whose vertices are the midpoints of the edges of the original. *The midpoint map* is a good name for this construction, and we consider it as a mapping on the space of polygons with a fixed number of vertices. When written in coordinates, the midpoint map is a linear transformation that is closely related to the heat equation: The vertex coordinates of the new polygon are averages of the vertex coordinates of the original. The midpoint map commutes with affine transformations of the plane. If you move the original polygon by an affine transformation, then the new one goes along for the ride.

In this monograph I will study a non-linear construction that is somewhat like the midpoint map but which commutes with projective transformations. I think of the construction as something like a cross between the midpoint map and the so-called pentagram map. I call the construction the *projective heat map* because I imagine – perhaps with scant justification – that the construction models how heat might flow in a world governed by projective geometry.

The projective heat map starts with a polygon P and returns a new polygon $P' = H(P)$. The figure illustrates the construction when the polygons involved are pentagons. P is the outer black one with white vertices, and P' is the inner black polygon with grey vertices. The auxiliary grey lines are just scaffolding for the construction.

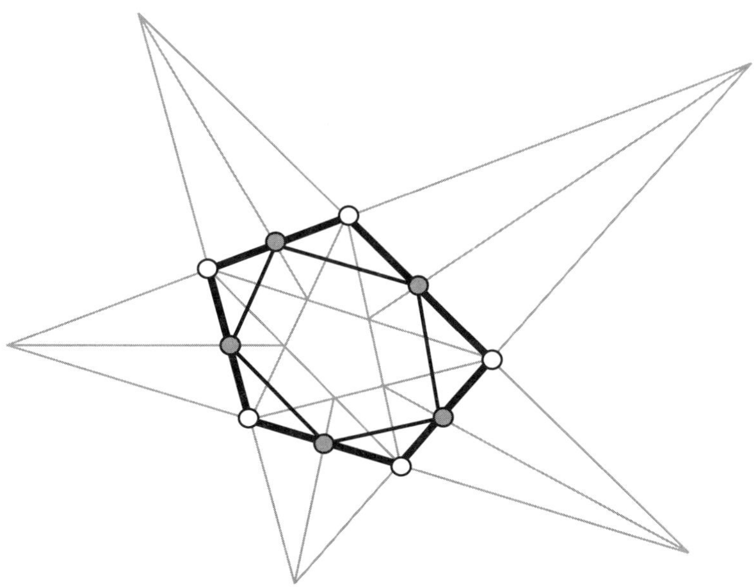

The main purpose of this monograph is to answer the question: *What does the projective heat map do to pentagons?* That is, what happens when the construction is iterated; what is the sequence $\{H^n(P)\}$ like? The question leads naturally to the study of a certain 2-dimensional real rational map. This rational map has surprisingly intricate behavior and a beautiful "Julia set". I will give rigorous, computer-assisted proofs of structural results which capture the main features of the projective heat map as it acts on pentagons, and this will give a pretty complete answer to the original question. I wrote an extensive graphical user interface which illustrates almost all the constructions of the monograph. This program is intended as a companion for the monograph.

To broaden the scope of this monograph, I will also discuss (in less depth) some other interesting polygon iterations such as the midpoint map, the map derived from Napoleon's Theorem, and the pentagram map. I will also include a numerical study of what the projective heat map does to N-gons in general, and how it interacts with the pentagram map.

This book is suitable for graduate students interested in projective geometry, rational maps, and polygon iterations. Most of the introductory chapters could also be read by advanced undergraduates. To make this work accessible to a broader audience, I have included some expository material on projective geometry, elementary algebraic geometry, and basic dynamical systems such as the one-sided shift and the Smale horseshoe. All these things play a role in the analysis of the projective heat map.

In his forthcoming 2017 Brown University Ph.D. thesis, Quang Nhat Le explores a 1-parameter family of maps generalizing the projective heat map. This monograph does not discuss these matters, but the interested reader might like to know about the existence of Nhat's thesis.

I thank Max Glick, Pat Hooper, Quang Nhat Le, Gloria Mari-Beffa, Curtis McMullen, Valentin Ovsienko, John Smillie, Sergei Tabachnikov, Guilio Tiozzo, and Amie Wilkinson for helpful and interesting discussions. I also thank an anonymous referee for very helpful and detailed suggestions and comments. Finally, I thank the National Science Foundation and the Simons Foundation for their support.

CHAPTER 1

Introduction

1.1. From Geometry to Dynamics

Figure 1.1 shows the constructions for three classical theorems in geometry.

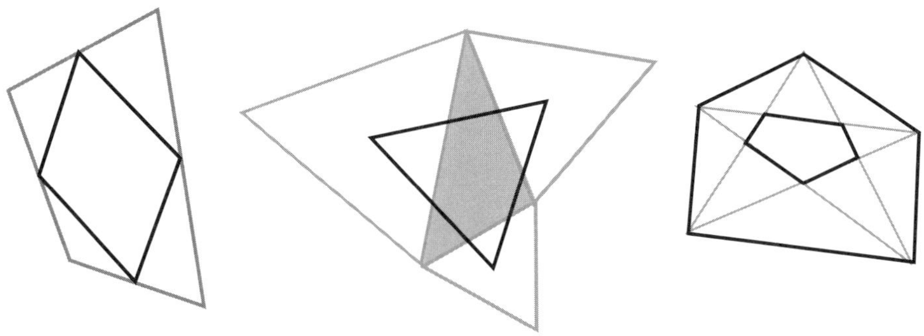

FIGURE 1.1. Three classic theorems

(1) Start with an arbitrary quadrilateral. Then the midpoints of the edges of the quadrilateral are the vertices of a parallelogram. I don't know the origins of this result.

(2) Start with an arbitrary triangle and construct equilateral triangles on each of the sides. Then the centers of the three equilateral triangles themselves make an equilateral triangle. This is known as Napoleon's Theorem, and it is attributed (perhaps not in all seriousness) to the famous emperor.

(3) Start with an arbitrary pentagon. Connect the vertices in a star pattern and consider the smaller pentagon in the middle. The inner pentagon and the outer pentagon are equivalent by a projective transformation. In some sense this result was known to Darboux, and it is studied explicitly in [**Mot**] and [**S1**].

All these results have probably been rediscovered many times. I will give proofs of the first two of these results in §2, and the third one in §5.1. (The first one is very easy.)

Figure 1.2 shows the same constructions for polygons having more vertices, and one can ask whether the theorems above generalize. Strictly speaking, the results above do not generalize as configuration theorems. For instance, the quadrilateral joining the centers of the equilateral triangles is not typically a square.

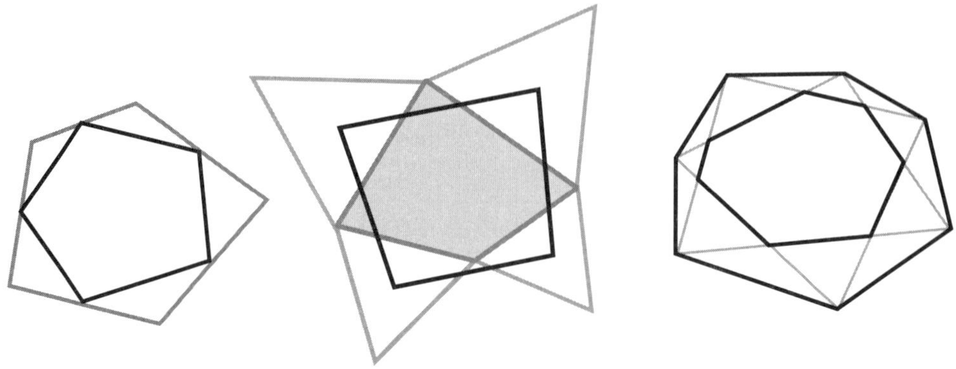

FIGURE 1.2. Variants of the classic constructions

However, there are dynamical generalizations of the results. For each of the constructions, let P be the initial polygon and let P' denote the polygon defined by the construction. Let $P^{(2)} = P''$ and $P^{(3)} = P'''$, etc.

In the first case, it is useful to work in $\mathcal{A}_N$, the space of N-gons modulo affine transformations. An *affine transformation* is the composition of a linear isomorphism and a translation. Two N-gons are equivalent if there is an affine transformation carrying one to the other. The midpoint map commutes with affine transformations, so it makes sense to consider $\{P^{(n)}\}$ as a sequence in $\mathcal{A}_n$. The general result is that $\{P_n\}$ converges to the equivalence class of the regular N-gon for almost all choices of P. This is a classical result which is closely related to heat flow and the discrete Fourier transform. I will give a proof in §2.

In the second case, it is useful to work in the space $\mathcal{S}_N$ of N-gons modulo similarity. Napoleon's construction commutes with similarities and so it makes sense to consider $\{P^{(n)}\}$ as a sequence in $\mathcal{S}_N$. What typically happens to this sequence depends on N. For example, when $N = 12$ the sequence $\{P^{(n)}\}$ converges to the class of the 12-gon which wraps twice around the regular hexagon for almost all choices of P. §2.4 I'll give an analysis of most cases. See also [**Bo**] and [**Gr**].

The third construction is the pentagram map, which I introduced in [**S1**]. In recent years, there have been many papers on the pentagram map. See the references listed in §5.2. The natural setting for the pentagram map is the space $\mathcal{P}_N$ of projective equivalence classes of N-gons. Here, the N-gons are considered subsets of the projective plane, and two N-gons are equivalent if there is a projective transformation carrying one to the other.

In brief, the projective plane $\boldsymbol{RP}^2$ is the space of lines through the origin in $\boldsymbol{R}^3$, and a projective transformation is a map on $\boldsymbol{RP}^2$ induced by an invertible linear transformation on $\boldsymbol{R}^3$. In §3 I will give a primer on projective geometry and explain this in more detail. The basic result here, due to V. Ovsienko, myself, and S. Tabachnikov, [**OST1**], [**OST2**], and independently due to F. Soloviev [**Sol**], is that the pentagram map is (in the appropriate sense) a discrete completely integrable system on $\mathcal{P}_N$. I will discuss this and other results about the pentagram map in §5.

All the results above make statements about *polygon iterations*. One starts with an N-gon P and produces a new N-gon P' by some geometric construction. One then asks about the behavior of the sequence $\{P^{(n)}\}$, perhaps modulo some group of symmetries. The main purpose of this monograph is to study a construction that is related to the ones above, which I call the *projective heat map*.

1.2. The Projective Heat Map

The projective heat map is based on the following construction. Given 4 points $a_{-3}, a_{-1}, a_1, a_3 \in \boldsymbol{RP}^2$ in general position – i.e. no three lie on a line – there is a canonical choice of a point b_0 on the line $\overline{a_{-1}a_1}$. The construction is shown in Figure 1.3. One might call b_0 the *projective midpoint* of a_{-3}, a_{-1}, a_1, a_3. Note that b_0 is typically not the actual midpoint of the segment $\overline{a_{-1}a_1}$. One situation where this does happen is when there is some isometry that swaps a_{-3} with a_3 and a_{-1} with a_1, but this is not the typical situation.

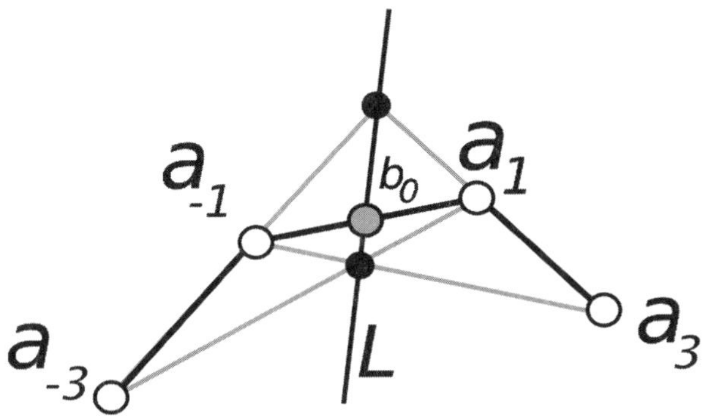

FIGURE 1.3. The construction of b_0 from a_{-3}, a_{-1}, a_1, a_3

Starting with an N-gon P, with vertices $\ldots a_{-3}, a_{-1}, a_1, a_3, \ldots$ we form the N-gon P' with vertices $\ldots b_{-2}, b_0, b_2, \ldots,$ where b_k is the projective midpoint of $a_{k-3}, a_{k-1}, a_{k+1}, a_{k+3}$. The indices are taken cyclically in this construction. We define $H(P) = P'$. The map H commutes with projective transformations and thus H induces a map on the quotient space $\mathcal{P}_N$ of projective equivalence classes of N-gons. The subspace $\mathcal{C}_N$ of equivalence classes of convex N-gons is an invariant subspace. It is worth remarking that H, like the pentagram map, is not entirely defined on $\mathcal{P}_N$. The points of the polygon need to be in sufficiently general position for this to make sense.

Numerical evidence supports the following conjecture.

CONJECTURE 1.1. *For any $N \geq 5$, and for almost all $P \in \mathcal{P}_N$, the sequence $\{P^{(n)}\}$ converges to the projectively regular class.*

I will discuss Conjecture 1.1 in §7.5. What makes this conjecture difficult is that, in contrast to the midpoint map and Napoleon's construction, the map H is nonlinear. Even in the case $N = 5$, where the basic space $\mathcal{P}_5$ is a 2 dimensional space, there is a lot of complexity. In the case $N = 5$, the projective heat map gives rise to a two-variable rational map

$$H(x, y) = (x', y'),$$

$$x' = \frac{(xy^2 + 2xy - 3)\,(x^2y^2 - 6xy - x + 6)}{(xy^2 + 4xy + x - y - 5)\,(x^2y^2 - 6xy - y + 6)}$$

$$y' = \frac{\left(x^2y+2xy-3\right)\left(x^2y^2-6xy-y+6\right)}{\left(x^2y+4xy-x+y-5\right)\left(x^2y^2-6xy-x+6\right)} \tag{1.1}$$

Our notation is such that H stands both for the map on $\mathcal{P}_5$ and the above rational map on $\boldsymbol{R}^2$.

1.3. A Picture of the Julia Set

The main goal of this monograph is to prove Conjecture 1.1 for the case $N = 5$. At this point we set $\mathcal{P} = \mathcal{P}_5$, etc., because this is the only case we consider. To be precise, $\mathcal{P}$ consists of projective equivalence classes of pentagons whose points are in general position.

We let $\mathcal{J}$ be the subset of $\mathcal{P}$ consisting of those points with well-defined orbits that do not converge to the regular class. The set $\mathcal{J}$ is akin to the Julia set from complex dynamics, a topic we discuss in §6.1. Our proof of Conjecture 1.1 amounts to showing that $\mathcal{J}$ has measure 0. However, given the beauty of $\mathcal{J}$, I couldn't resist analyzing it and getting finer information about it. Almost all my motivation for this monograph came from wanting to rigorously justify the computer pictures of $\mathcal{J}$ which I produced.

Figure 1.4 shows the most interesting portion of $T \circ \boldsymbol{B}(\mathcal{J})$, where T is a certain linear transformation and $\boldsymbol{B}$ is a birational map discussed below. The points are colored according to how many iterates of H it takes to map them into $\mathcal{C}$, the space of convex classes, and then these colored points are mapped into the picture plane *via* $T \circ \boldsymbol{B}$. Once a point gets into $\mathcal{C}$ it converges under iteration to the regular class. (See Theorem 1.3 below.) The darker the color, the longer it takes. So, $T \circ \boldsymbol{B}(\mathcal{J})$ would be the black points – or at least the black points with well defined orbits.

1.4. The Core Results

Our first result is topological in nature.

THEOREM 1.2. *The map H is generically 6-to-1 when it acts on $\boldsymbol{C}^2$.*

Our next result deals with the action of H on $\mathcal{C}$.

THEOREM 1.3. *There is a smooth and rational function $f : \mathcal{C} \to \boldsymbol{R}$ such that $f \circ H(P) \geq f(P)$ for all $P \in \mathcal{C}$, with equality if and only if P is the regular class. Moreover, the level sets of f are compact.*

Given that f has compact level sets, Theorem 1.3 has the following corollary. See the proof of Corollary 9.7 for details.

COROLLARY 1.4. *$\{H^n(P)\}$ converges to the regular class for all $P \in \mathcal{C}$.*

Now we turn to Conjecture 1.1. The way we understand $\mathcal{J}$ is that we first understand a certain Cantor set in $\mathcal{J}$ and then we understand the rest. Accordingly, here are our two main structural results.

THEOREM 1.5. *$\mathcal{J}$ contains a measure 0 forward-invariant Cantor set $\mathcal{J}C$. The restriction of H to $\mathcal{J}C$ is conjugate to the 1-sided shift on 6-symbols.*

THEOREM 1.6. *$\mathcal{J}$ contains a measure 0 forward-invariant Cantor band $\mathcal{J}A$ such that*

$$\mathcal{J} = \mathcal{J}C \cup \bigcup_{k=0}^{\infty} H^{-k}(\mathcal{J}A).$$

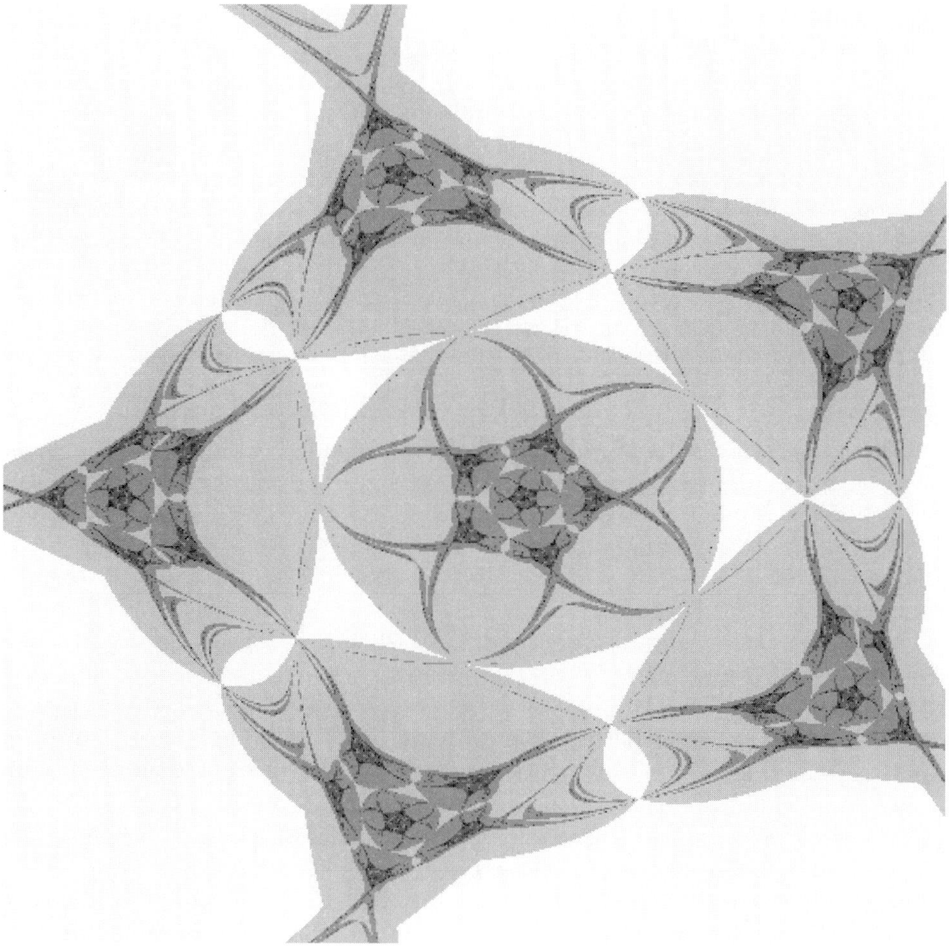

FIGURE 1.4. A subset of $\mathcal{J}$.

$\mathcal{J}A$ is an open subset of $\mathcal{J}$ in the subspace topology. The action of H in a neighborhood of $\mathcal{J}A$ is the 10*-fold covering of a quasi-horseshoe.*

Remarks:
(i) A *Cantor band* is a space homeomorphic to the product of a Cantor set and an open interval. If you look at Figure 1.4 you can see a lot of Cantor bands.
(ii) We discuss the one sided shift in §6.2.
(iii) In §6.5 we define what we mean by a quasi-horseshoe. Such maps are pretty close to the Smale horseshoe, which we describe in §6.4. When we prove theorem 1.6 we will explain what we mean by a 10-fold covering.

COROLLARY 1.7. *Conjecture* 1.1 *holds for $N = 5$.*

Proof: $\mathcal{P}$ inherits a smooth structure from its inclusion into $\boldsymbol{R}^2$. Recall that a smooth map is *regular* at p if it is a local diffeomorphism at p. Almost every point in $\mathcal{P}$ has a well defined H-orbit in which every power of H is regular at p. (The set of points which do not have this property is contained in a countable union of algebraic curves.) Call such points *totally regular*.

Let $\mathcal{X} = \mathcal{J} - \mathcal{J}C$ and let $\mathcal{Y}$ denote the set of totally regular points in $\mathcal{X}$. Since almost every point is totally regular, and since $\mathcal{J}C$ has measure 0, it suffices to prove that $\mathcal{Y}$ has measure 0.

Call an open disk Δ *clean* if there is some k such that the restriction $H^k|_\Delta$ is a diffeomorphism and if $H^k(\Delta \cap \mathcal{X}) \subset \mathcal{J}A$. In this case $\Delta \cap \mathcal{X}$ is the image of a subset of $\mathcal{J}A$ under a diffeomorphism, and hence has measure 0. If follows from Theorem 1.6 that $\mathcal{Y}$ is covered by a countable union of clean disks. Hence $\mathcal{Y}$ is the countable union of sets of measure 0. Hence $\mathcal{Y}$ has measure 0. $\square$

1.5. Deeper Structure

For most of the results above, and all the results in this section, we will instead use the map $\boldsymbol{H} = \boldsymbol{B}H\boldsymbol{B}^{-1}$, where

$$\boldsymbol{B}(x,y) = (b(x), b(y)), \qquad b(t) = \phi^3\left(\frac{\phi + t}{-1 + \phi t}\right) \tag{1.2}$$

Here $\phi = (1+\sqrt{5})/2$ is the golden ratio.

After we change coordinates, we will replace the space $\boldsymbol{R}^2$ by the more global space $\boldsymbol{M}$ obtained by blowing up $(\boldsymbol{R} \cup \infty)^2$ at 3 specially chosen points. The space $\boldsymbol{M}$ is a nice compact moduli space of projective classes of labeled pentagons – not necessarily in general position. It turns out that there is an order 10 group $\boldsymbol{\Gamma}$ of birational diffeomorphisms of $\boldsymbol{M}$ corresponding to dihedral relabelings of the pentagons. Beautifully, there is a fundamental domain for $\boldsymbol{\Gamma}$ acting on $\boldsymbol{M}$ that is obtained by blowing up one corner of a particular Euclidean triangle. This fundamental domain turns out to be extremely useful to us.

Another advantage of using $\boldsymbol{H}$ and $\boldsymbol{M}$ is that the fixed points of $\boldsymbol{H}$ in $\boldsymbol{M}$ have a particularly nice form. One fixed point is (∞, ∞), corresponding to the regular class. Another fixed point is $(0,0)$, corresponding to the star regular [1] class. Finally, there are 5 additional fixed points corresponding to the $\boldsymbol{\Gamma}$ orbit of $(1,1)$. These 5 points represent various labelings of the star convex pentagon shown in Figure 7.4. I discovered the map $\boldsymbol{B}$ by trying to move the fixed points of H to the nicest possible locations, and then the triangular fundamental domain turned out to be a happy surprise.

Figure 1.4 really shows the Julia set for $\boldsymbol{H}$, up to a linear transformation [2] which is chosen so that the differentials of elements in $\boldsymbol{\Gamma}$ act isometrically at $(0,0)$.

We let $\heartsuit J$ denote [3] the closure of the set of points in $\boldsymbol{M}$ which have well defined $\boldsymbol{H}$-orbits but which do not converge to (∞, ∞). We think of $\heartsuit J$ as a kind of completion of $\mathcal{J}$.

We say that a *cone point* is a point in $\heartsuit J$ having arbitrarily small neighborhoods which intersect $\heartsuit J$ in the cone on a Cantor set. Intuitively, the cone points are

---

[1] A pentagon is *star regular* if the relabeling relabeling $(1,2,3,4,5) \to (1,3,5,2,4)$ makes it regular, and *star convex* if this relabeling makes it convex.

[2] This linear transformation makes the geometric picture as nice as possible, but we don't use it in our analysis because it is defined over a fairly high degree number field.

[3] We put the symbol $\heartsuit$ in front of subsets of the Julia set in $\boldsymbol{M}$ to avoid notational clashes what the many other objects that get letter names.

where the Cantor bands pinch down to single points. See Figure 1.4. Here is a general structural result.

THEOREM 1.8. *$\heartsuit J$ is the union of a Cantor set $\heartsuit JC$, a countable collection of Cantor bands, and a countable collection of cone points.*

Here $\heartsuit JC = \boldsymbol{B}(\mathcal{J}C)$, where $\mathcal{J}C$ is as in Theorem 1.5.

Now we describe structures in $\heartsuit J$ which elaborate the ones from Theorems 1.5 and 1.6. Consider the following infinite graph. One starts with the finite graph shown on the left side of Figure 1.5 below and then puts the same graph inside each of the 6 shaded pentagonal "holes". This produces a more complicated graph with 36 pentagonal holes, shown on the right hand side of Figure 1.5. One then repeats indefinitely. Call the limiting "graph" $\mathcal{G}_\infty$. This space is a variant of the Sierpinski triangle.

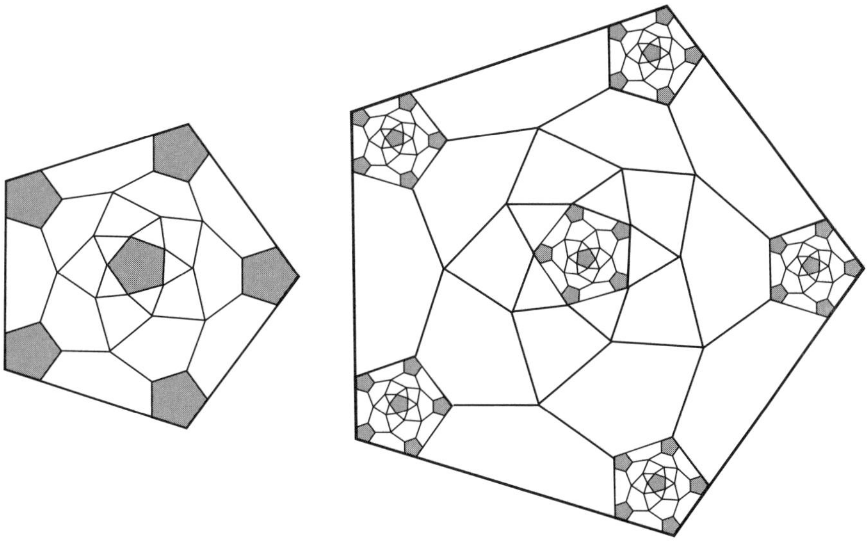

FIGURE 1.5. The seed for $\mathcal{G}_\infty$ and the second step in the construction.

THEOREM 1.9. *$\heartsuit J$ contains a subset $\heartsuit G$ which is homeomorphic to $\mathcal{G}_\infty$.*

THEOREM 1.10. *$\heartsuit J$ contains a forward invariant subset $\heartsuit S$ which, when blown up at all its cone points. is homeomorphic to the connected 5-fold cover of the 2-adic solenoid.*

Remarks:

(i) $\heartsuit JC$ is the subset of $\heartsuit G$ comprised of the nested intersections of pentagonal holes in the iterated construction.

(ii) In local coordinates, $\heartsuit G$ is a compact subset of $\boldsymbol{R}^2$. The filled-in version Fill$\heartsuit G$, i.e. the complement of the unbounded component of $\boldsymbol{R}^2 - \heartsuit G$, is a "solid pentagon" whose 5 "vertices" are the orbit $\boldsymbol{\Gamma}(1,1)$. The center of $\heartsuit G$ is $(0,0)$.

(iii) The connected 5-fold cover of the 2-adic solenoid is the quotient

$$(\boldsymbol{R} \times \boldsymbol{Z}_2)/\sim, \qquad (x,y) \sim (x+5n, y+5n), \qquad n \in \boldsymbol{Z}.$$

Here $\boldsymbol{Z}_2$ is the topological group of 2-adic integers. We will discuss this space in more detail in §6.6.
(iv) $\heartsuit S$ is the closure of the union of the maximal C^1 arcs of $\heartsuit J$ which intersect the Cantor band $\heartsuit JA = \boldsymbol{B}(\mathcal{J}A)$. Here $\mathcal{J}A$ is the Cantor band from Theorem 1.6.
(v) Our last picture in the monograph, Figure 20.7, is the culmination of all our analysis. It shows a detailed schematic picture of $\heartsuit G$ and $\heartsuit S$ sitting inside $\boldsymbol{M}$. We get the more precise result that $\heartsuit S = (\heartsuit J - \text{Fill}\heartsuit G) \cup \boldsymbol{\Gamma}(1,1)$.

All this structure contributes to our final result:

THEOREM 1.11. *The Julia set $\heartsuit J$ is path connected.*

Note that $\heartsuit J$ is obviously not locally connected, on account of all the Cantor bands. So, the connectivity comes about in a complicated way.

1.6. A Few Corollaries

The Cantor set $\mathcal{J}C$ from Theorem 1.5 is contained entirely within the set of star convex classes, and the one-sided shift has periodic points of all orders. Hence H has periodic points of all orders, and we can even find such periodic points where every point in the orbit represents a star convex pentagon.

For the interested dynamics expert, I will sketch a proof in §15.6 that H is not post-critically finite. That is, the forward images of the set where dH is singular cannot be contained in a finite union of algebraic curves. What happens is that the singular set gets mapped transversely across the quasi-horseshoe and then it gets wrapped around like crazy, making it impossible for the forward image to be contained in a finite union of algebraic curves. If H were post-critically finite, there would be additional tools available to investigate H, as in [**N**], so the result here rules out one possible shortcut to the analysis of H.

Also for the dynamics expert, I will sketch a proof in §15.7 that H is not rationally conjugate to a one-dimensional rational map. That is, there is no pair (f,h) where $f: \boldsymbol{R}^2 \to \boldsymbol{R}$ and $h: \boldsymbol{R} \to \boldsymbol{R}$ are rational and $fH = hf$. This situation is impossible because some fibers of f would either cross the horseshoe transversely or run along the leaves of the solenoid. Either case leads to a contradiction. The lack of a rational semi-conjugacy rules out another possible shortcut to the analysis of H.

1.7. Sketch of the Proofs

Theorem 1.2 has an elementary algebraic geometry proof. We count roots of an associated pair of polynomials using Bezout's theorem.

Theorem 1.3 is a direct calculation once we identify the increasing quantity. The increasing quantity turns out to be $E_5 O_5$, the simplest of the pentagram map invariants. See §5.3.

For the remaining results, we partition $\mathcal{P}$ into a finite union of polygonal pieces on which the action on the pieces is simple enough to analyze. Our partition will have roughly the following structure.

- Some pieces will map into $\mathcal{C}$ after finitely many steps.
- Some pieces will map over themselves in an expanding way. This will give rise to the Cantor set $\mathcal{J}C$ from Theorem 1.5.

- Some pieces will map over themselves in a hyperbolic way. Roughly, they will be stretched in one direction and contracted in another. This will give rise to the quasi-horseshoe $\mathcal{J}A$ from Theorem 1.6.

The proofs of the results mentioned in §1.5 build on the properties of $\mathcal{J}C$ and $\mathcal{J}A$ and also make use of our partition.

The novel part of our approach is how we rigorously prove that pieces in the partition move as we think that they do. Essentially, we boil down every step to verifying that the image of some (solid) polygonal subset of $\mathcal{P}$ under f is contained in, or disjoint from, another (solid) polygonal subset. We reduce both questions to statements that certain finite collections of polynomials are positive (or non-negative) on certain finite collections of polygons. We then use a divide-and-conquer algorithm to establish this positivity. We call our algorithm *the method of positive dominance*. We explain it in §11.

Everything involved in our construction is defined over the ring $\boldsymbol{Z}[1/2, \sqrt{5}, \sqrt{13}]$, and so all our calculations are exact integer calculations. There are no round-off errors. Our results sometimes require the analysis of polynomials having total degree about 45 and coefficients whose integer components are about 50 digits long. To handle the enormous polynomials we get, we implement our calculations in Java, using the BigInteger class. The BigInteger class allows one to do arithmetic involving integers which are thousands of digits long. Our calculations don't come anywhere near the size limit imposed by the hardware of modern computers.

1.8. Some Comparisons

In spite of the complexity of the equation for the projective heat map, the results we get for this real variable map on $\boldsymbol{R}^2$ are almost comparable in detail to the kinds one sees in one dimensional complex dynamics. See [**Mil**] for an introduction to that vast subject. For instance, our Theorem 1.9 is similar in spirit to the combinatorial models of Julia sets called *Hubbard trees*. In §6.1 we will discuss Julia sets of one dimensional rational maps and compare them to our $\mathcal{J}$.

The projective heat map is defined over almost any field, and in particular makes sense on the complexified version of $\mathcal{P}$. Thus, one could consider the projective heat map in the context of 2-variable complex dynamics. There is a large literature on rational maps on $\boldsymbol{C}^2$ or on other complex surfaces. For instance, the paper [**BLS**] is one of a long series of papers written by the authors on the case of polynomial maps. The projective heat map is not a polynomial map, so works like [**BLS**] would probably not help with the proofs with the results above, but they might be a beacon for future research.

Our quasi-horseshoe result is akin to results about the Hénon map:

$$F_{a,b}(x, y) = (1 - ax^2 + y, by). \tag{1.3}$$

Here a and b are parameters which influence the nature of the map. The classic case is $a = 3/10$ and $b = 14/10$. In this case, there is an attracting Cantor band. As long as $a \neq 0$, the Hénon map is a polynomial diffeomorphism of $\boldsymbol{R}^2$. The papers [**A1**] and [**A2**] deal with computational techniques for finding subsets of parameters (a, b) for which $F_{a,b}$ acts as the Smale horseshoe on the set of bounded orbits. These techniques are somewhat like ours except that they use interval arithmetic and they deal with an entire family of maps. The paper also [**BS**] has a discussion of this problem.

In terms of the method of proof, one could also compare our results to those in [**Tu**] concerning the existence of the Lorenz attractor. This paper also uses a kind of partition approach to deal with a single dynamical system.

While not giving a comprehensive list, let me mention some other papers which get detailed dynamical pictures for real rational maps of the plane. The results in these papers, while certainly inspired by computer experiments, involve traditional proofs. One thing about all these other papers is that the formulas for the maps are considerably simpler than the formula for the projective heat map. Either the high degree nature of the projective heat map makes it too difficult to study in a traditional way – witness the difference in the amount known about the dynamics of quadratic polynomials and the dynamics of higher degree polynomials – or else a smarter author could do a better job.

The paper [**BD**] studies the map $\tau \circ \sigma$, where

$$\sigma(x,y) = \left(1 - x + \frac{x}{y}, 1 - y + \frac{y}{x}\right), \qquad \tau(x,y) = (x, bx + a + 1 - y). \tag{1.4}$$

This is a 2-parameter family of bi-rational maps depending on parameters a and b.

The paper [**BLR**] deals with the single rational map coming from (a case of) the Migdal Kadanoff RG equations

$$(x,y) \to \left(\frac{x^2+y^2}{x^{-2}+y^2}, \frac{x^2+x^{-2}+2}{x^2+x^{-2}+y^2+y^{-2}}\right). \tag{1.5}$$

These equations have to do with the Ising model.

The paper [**N**] studies the complex dynamics of the map

$$(z,w) \to \left((1-2z/w)^2, (1-2/w)^2\right) \tag{1.6}$$

and constructs a combinatorial model for the Julia set. Our Theorem 1.9 is similar in spirit to this, though we get less information. In this map, the forward orbit of the critical set – i.e. where the map is not a local diffeomorphism – is just a finite union of lines. The map is an example of a post-critically finite map, unlike the projective heat map.

The paper [**HP**] studies the dynamics of Newton's method when it is used to solve two simultaneous quadratic equations. (It is somewhat difficult to extract as concrete formula as the ones given above.) This leads to a rational map on $\boldsymbol{C}^2$ which the authors analyze in detail.

Finally, the paper [**BDM**] also detailed information about the structure of some concrete rational self-maps of $\boldsymbol{CP}^2$.

1.9. Outline of the Monograph

This monograph comes in 3 parts.

Part 1: Context: Here I place the projective heat map in a broader context. I include some background on projective geometry and dynamical system, and analyze some dynamical systems related to the projective heat map. Some readers might like §5, which has an account of some of the main features of the pentagram map, including integrability. The sections directly relevant to the projective heat map are §3, §4.1-4.3, §5.3, §5.6, §6.2 and §6.5.

Part 2: The Core Results: In this part of the monograph, I prove Theorems 1.2, 1.3, 1.5, and 1.6. I also explain the method of positive dominance, which is used throughout this part and Part 3. This is the core material. At the end of Part 2, the proof of the pentagon case of Conjecture 1.1 is done.

Part 3: Deeper Structure: This part of the monograph studies the structure of $\mathcal{J}$ in more detail. In particular, I prove Theorems 1.11, 1.8 and 1.9 and 1.10. (This is not the order in which the results are proved.) The arguments in this part are considerably more intricate than the ones in Part 2. To help guide the reader through the thicket of details, I have included an introductory chapter, §15, which gives detailed sketches of all the proofs.

At the end of the monograph, I have included a reference chapter in an attempt to make this monograph easier to read. §21.1 lists many of the basic definitions and objects in the monograph and points to where they are discussed. The rest of the chapter lists out important formulas, including coordinates for the vertices of the polygons in our partition. We mention §21.3 especially. This section has the formulas for all our basic maps.

1.10. Companion Program

I discovered practically everything in this monograph by writing a java computer program which implements the dynamics of the projective heat map. The computer-assisted part of my proof resides in the program. The reader can launch the proofs using the program, and survey them down to a fine level of detail.

I strongly encourage the reader of this monograph to download and use the program. The heavily documented program illustrates practically everything about our results, and also has a tutorial section which teaches the user how to run the computational tests used in our proofs. I have tried to write the monograph so that it stands on its own, but I think that the reader will have a much more satisfying experience reading the monograph while operating the program and seeing vivid illustrations of what is going on. I would say that this program relates to the monograph much like a movie relates to its screenplay.

One can download the program from

http://www.math.brown.edu/~res/Java/HEAT2.tar

The program has a README file which explains how to compile and run the program.

Part 1

CHAPTER 2

Some Other Polygon Iterations

In this chapter, we analyze the two iterations mentioned at the beginning of the introduction, the midpoint map and the one connected to Napoleon's Theorem. We also discuss a third iteration that is based on conformal geometry.

2.1. The Midpoint Theorem

Let P be a quadrilateral and let P' be the new quadrilateral whose vertices are the midpoints of P. Let D be one of the diagonals of P.

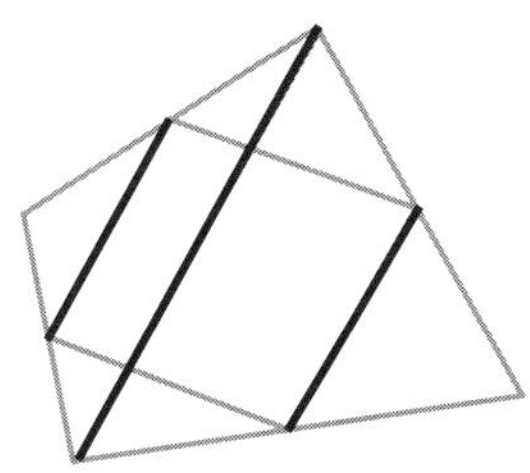

FIGURE 2.1. Proof of the parallelogram result

Using similar triangles, you can check that two of the sides of P' are parallel to D and hence to each other. Since this holds for each diagonal of P, we see that the opposite sides of P' are parallel in pairs. That is, P' is a parallelogram.

2.2. The Midpoint Iteration

We define the midpoint construction on the set X_N of oriented N-gons, normalized so that the center of mass of the vertices is the origin. That is, X_N is the complex vector subspace of $\boldsymbol{C}^N$ consisting of vectors $(z_1, \ldots, z_N)$ such that $z_1 + \cdots + z_N = 0$. The midpoint map is given by $M(z_1, \ldots, z_N) = (z'_1, \ldots, z'_N)$, where

$$z'_k = \frac{1}{2} z_k + \frac{1}{2} z_{k+1}. \tag{2.1}$$

In making this definition, we have made a symmetry-breaking choice on how to label the vertices of the new polygon, but this choice turns out to be irrelevant for the final analysis.

The most significant fact is that the map $M : X_N \to X_N$ is a circulent linear map. (Here *circulent* means that M commutes with the cyclic shifting of the coordinates.) Since M is linear, it makes sense to look for a basis of eigenvectors. Since M is circulent, the eigenvectors have a very specific form.

Let $\omega_N = \exp(2\pi i/N)$. The $n-1$ dimensional complex vector space X_N has the basis $E_1, \ldots, E_{N-1}$, where

$$E_k = (1, \omega_N^k, \omega_N^{2k}, \ldots, \omega_N^{(N-1)k}). \tag{2.2}$$

It follows from symmetry that $E_1, \ldots, E_{N_1}$ are all eigenvectors for M. Let $\lambda_1, \ldots, \lambda_{n-1}$ be the corresponding eigenvalues. We have

$$|\lambda_k| = |\lambda_{N-k}| = \frac{|1+\omega^k|}{2}. \tag{2.3}$$

From this equation, we see that

$$|\lambda_1| = |\lambda_{N-1}| > |\lambda_2| = |\lambda_{N-2}| > |\lambda_3| = |\lambda_{N-3}| \cdots \tag{2.4}$$

Every $P \in X_N$ can be written

$$P = \sum_{k=1}^{N-1} a_k E_k. \tag{2.5}$$

Almost every choice will have

$$a_1 \neq 0, \qquad a_{N-1} \neq 0, \qquad |a_1| \neq |a_{N-1}|.$$

We have

$$M^n(P) = \sum_{k=1}^{N-1} \lambda^k a_k E_k = \mu_n \Big(E_1 + b_n E_{N-1} + \epsilon_n \Big) \tag{2.6}$$

Here ϵ_n is some vector whose norm tends to 0 as $n \to \infty$ and $|b_n|$ is independent of n, and μ_n is some scaling factor.

Suppose first we mod out by similarities, so that $P \sim \mu P$ for any nonzero complex μ. Modulo similarities, we see that, on a subsequence, $M^n(P)$ converges to some linear combination $E_1 + bE_{N-1}$ with $b \neq 0$. Note that b depends on the subsequence we take, and this is why it is not really completely satisfactory to mod out by similarities. However, since $|\lambda_1| = |\lambda_{N-1}|$, we see that the norm $|b|$ does not depend on the subsequence. We have

$$|b| = |b_n| = |a_{N-1}/a_1| \neq 1.$$

To recognize $E_1 + bE_{n-1}$ as something familiar, we identify $\boldsymbol{C}$ with $\boldsymbol{R}^2$, and observe that the kth vertex is

$$\begin{bmatrix} 1-b_1 & b_2 \\ b_2 & 1-b_1 \end{bmatrix} \begin{bmatrix} \cos(2\pi k/n) \\ \sin(2\pi k/n) \end{bmatrix}. \tag{2.7}$$

Here $b = b_1 + ib_2$. In short, $E_1 + bE_{N-1}$ is the image of the regular N-gon under some linear transformation R_b. Note that $\det(R_b) = 1 - |b|^2$, so that R_b is invertible if and only if $|b| \neq 1$. Hence $T_1 + bT_{N-1}$ is the image of the regular N-gon under an invertible linear transformation.

Recall that $\mathcal{A}_N$ is the space of N-gons modulo affine transformations. Remembering our normalization that $\sum z_k = 0$, we observe that what we have proved is equivalent to the statement that $\{M^n(P)\}$ converges in $\mathcal{A}_n$ to the affinely regular class for almost every initial choice of $P \in \mathcal{A}_N$.

Remark: When $|b| \neq 1$, the polygon $E_1 + bE_{N-1}$ is inscribed in an ellipse whose shape only depends on $|b|$. Thus, in a certain sense, we really do get convergence when we just mod out by similarities.

2.3. Napoleon's Theorem

In Figure 1.1, Napoleon's construction [1] appears to be defined for triangles without regard for orientation, but to bring the analysis into the realm of linear algebra, as we did for the midpoint map, we give a definition that depends on the orientation of the triangle.

Suppose that the vertices of the triangle P are (z_1, z_2, z_3). We define points (w_1, w_2, w_3) by the requiring that $z_{k+1}w_kz_{k-1}$ make the vertices of an equilateral triangle that is oriented counterclockwise. Here the indices are taken cyclically.

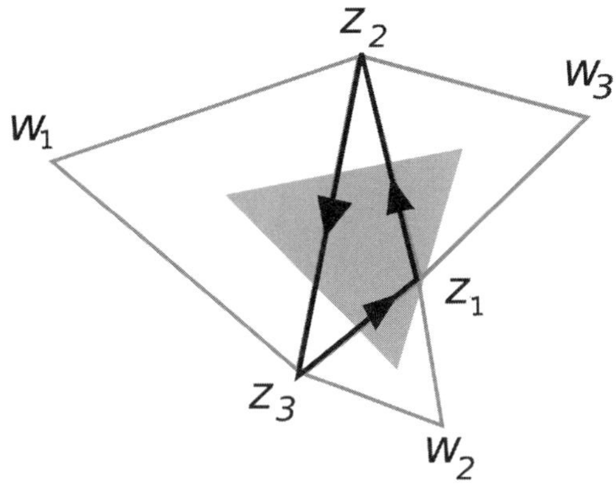

FIGURE 2.2. Napoleon's construction for oriented triangles.

Concentrating on w_1, we have

$$w_1 - z_2 = \omega(z_2 - z_3), \qquad \omega = \exp(2\pi i/3).$$

Solving this equation gives

$$w_1 = \omega(z_2 - z_3) + z_2.$$

One of vertices of the triangle P' given by Napoleon's construction is

$$z_1' = \frac{w_1 + z_2 + z_3}{3} = \frac{(2+\omega)}{3}z_2 + \frac{(1-\omega)}{3}z_3. \tag{2.8}$$

The formulas for the other two vertices are obtained by shifting the indices cyclically. The map $\Psi(P) = P'$ is again a circulent linear transformation in these coordinates.

As for the midpoint map, let X_3 denote the subspace of $\boldsymbol{C}^3$ consisting of points (z_1, z_2, z_3) with $z_1 + z_2 + z_3 = 0$. Let E_1 and E_2 be the basis for the space X_3 considered for the midpoint map. These vectors both represent equilateral triangles, with T_1 being oriented counterclockwise and T_2 being oriented clockwise.

Again, both E_1 and E_2 are eigenvectors for Ψ. This time the eigenvalues are $\lambda_1 = -1$ and $\lambda_2 = 0$. One can check this from the formula, but it is better just to draw a picture. Writing an arbitrary triangle as $P = a_1T_1 + a_2T_2$, we see by linearity that $P' = a_1T_1$. Hence P' is an equilateral triangle unless $P = a_2T_2$. In this case, P' is a single point. The unoriented construction given in Figure 1.1 implicitly assumes that P is oriented counterclockwise, and this gives us $a_2 \neq 0$. This is why the unoriented construction always produces a nontrivial triangle.

[1] Some authors doubt that this construction is truly due to Napoleon. See [**Gr**].

2.4. Napoleon's Iteration

Many generalizations of Napoleon's Theorem have been worked out – e.g. the Douglas-Neumann Theorem. See B. Grünbaum's article [**Gr**] or A. Bogomolny's online discussion [**Bo**] for plenty of information. Here I'll pursue one direction, without making any claims to the originality or completeness of the investigation.

As in the midpoint map, let X_N denote the vector space of N-gons $(z_1, \ldots, z_n)$ normalized so that $\sum z_i = 0$. Inspired by equation (2.8), we define $\Psi : X_N \to X_N$ by the formula $\Psi(z_1, \ldots, z_n) = (z_1', \ldots, z_n')$, where

$$z_k' = \frac{(2+\omega_3)}{3} z_{k+1} + \frac{(1-\omega_3)}{3} z_{k+2}. \tag{2.9}$$

Here $\omega_3 = \exp(2\pi i/3)$. This definition makes some arbitrary choices for the indices, but this does not bother us. Geometrically, the map T represents Napoleon's construction on oriented N-gons. The map Ψ is again a linear map, and again we consider the effect on the vector $E_1, \ldots, E_{N-1}$, which are all eigenvectors. Let $\lambda_1, \ldots, \lambda_{n-1}$ be the corresponding eigenvalues.

LEMMA 2.1. *The eigenvalues λ_k with the largest norm corresponds to the values of k which are closest to $N/6$. This value of k is unique when $N \not\equiv 3$ mod 6. When $N > 3$ and $N \equiv 3$ mod 6 there are two consecutive values of k which are closest to $N/6$.*

Proof: Let $z = \exp(i\theta)$ be a unit complex number. Let T_z be the equilateral triangle whose vertices are $1, w, z$, traced out in counterclockwise order as in Figure 2.3 below. Let z' be the center of T_z.

We study the function $h(\theta) = |z_\theta'|$. It is convenient to think of this as a function of $\theta \in [-2\pi/3, 4\pi/3]$. An easy exercise in calculus establishes the following

- h attains its global maximum, $2/\sqrt{3}$, at the midpoint $\pi/3$.
- h attains its minimum, 0, at the endpoints $-2\pi/3$ and $4\pi/3$.
- h is increasing on $[-2\pi/3, \pi/3]$ and decreasing on $[\pi/3, 4\pi/3]$.
- $h(\pi/3 - \theta) = h(\pi/3 + \theta)$ for all θ.

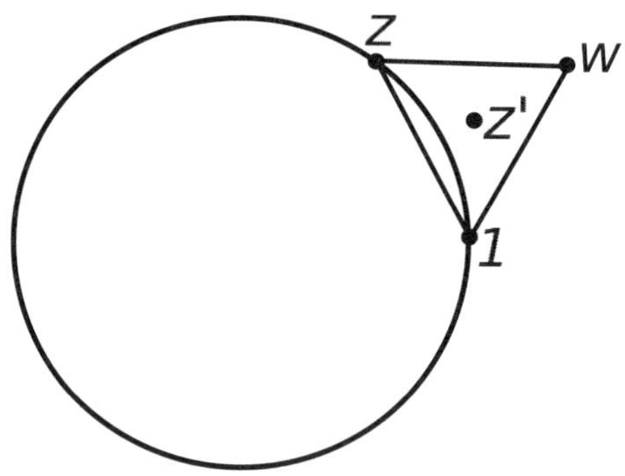

FIGURE 2.3. A point which varies with θ.

Let $\omega = \exp(2\pi i/N)$, as in the definition of the vectors $E_1, \ldots, E_{N-1}$. Let $\theta_k = 2\pi k/N$ be the argument of ω^k. We have

$$|\lambda_j| = h(\theta_j), \qquad j = 1, \ldots, N-1. \tag{2.10}$$

When $N \not\equiv 3 \bmod 6$, there is a unique choice of k which is closest to $N/6$ and the corresponding value of $\theta_k \in I$ is closest to $\pi/3$. Hence $|\lambda_k| > |\lambda_j|$ for any other $j \neq k$. When $N > 3$ and $N \equiv 3 \bmod 6$ there are two consecutive choices of k which are as close as possible to $N/6$, and we have $|\lambda_k| > |\lambda_j|$ when j is not one of these two special values of k. □

Now we play the same game as above for the midpoint map. For simplicity suppose that $N \not\equiv 3 \bmod 6$. We take an arbitrary $P \in X_N$ and write

$$P = \sum_{i=1}^{N-1} a_i E_i.$$

For almost every choice of p we will have $a_k \neq 0$, where k is the index closest to $N/6$. By Lemma 2.1 we have

$$\Psi^n(P) = \mu_n\Big(E_k + \epsilon_n\Big). \tag{2.11}$$

Here ϵ_n is a vector whose norm tends to 0 with n and μ_n is a scale factor. Modulo similarities, the sequence $\{P_n\}$ converges to E_k. When N is divisible by 6, the polygon E_k traces around the regular hexagon $N/6$ times. One could say that when N is not divisible by 3, the limiting shape tries as hard as possible to wrap itself around the regular hexagon. For $N = 4, 5, 6, 7, 8$ the limiting shape is the convex regular N-gon. For $N = 10$ the limiting shape wraps twice around the convex regular pentagon. Figure 2.4 shows the limiting shape when $N = 11$.

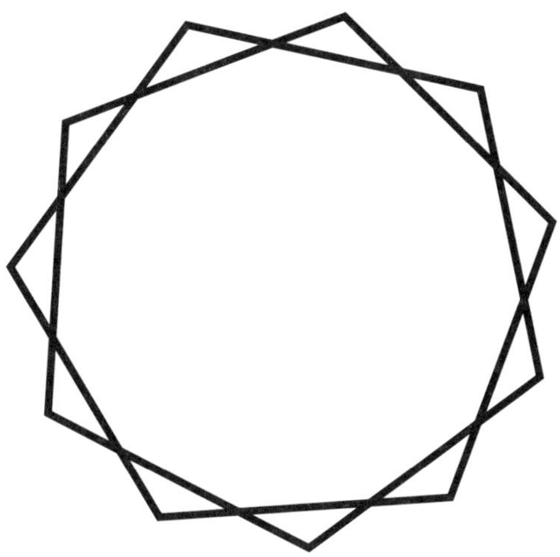

FIGURE 2.4. The limiting shape for $N = 11$.

When $N > 3$ and $N \equiv 3 \bmod 6$, the generic behavior is somewhat more subtle. We leave this to the interested reader.

2.5. Conformal Averaging

In [**S4**] I studied a polygon iteration that is similar in spirit to the projective heat map. I called this other iteration *conformal averaging*. The iteration works for convex polygons inscribed in the unit circle.

Given 4 complex numbers a, b, c, d, we define their *cross ratio*.

$$[a, b, c, d] = \frac{(a-b)(c-d)}{(a-c)(b-d)} \tag{2.12}$$

In the next chapter we will discuss the cross ratio in the alternate setting of real projective geometry. Here we are interested in the cross ratio of unit complex numbers. When a, b, c, d are unit complex numbers, their cross ratio is a real number.

Given 4 consecutive unit complex numbers a_{-3}, a_{-1}, a_1, a_3 on the unit circle, there is a unique unit complex number c_0 on the unit circle such that

$$[a_{-3}, a_{-1}, c_0, a_1] = [a_{-1}, c_0, a_1, a_3] \tag{2.13}$$

and the points $a_{-3}, a_{-1}, c_0, a_1, a_3$ come in order on the unit circle. This equation has two solutions, and we take the one which makes the points lie in the correct order.

The left side of Figure 2.5 shows a geometric construction of c_0. For comparison, the right side shows the construction of the projective midpoint b_0. The a points are in black and move from left to right.

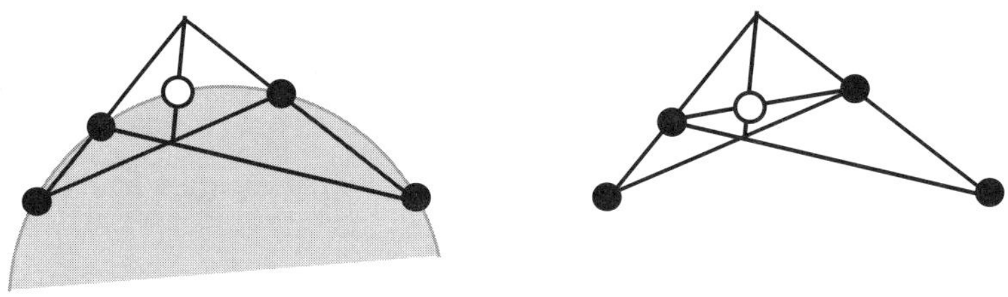

FIGURE 2.5. The conformal and projective midpoints

We think of an *inscribed convex N-gon* as a cyclically ordered list

$$P = (a_1, a_3, a_5, \dots)$$

of N unit complex numbers. Geometrically P is a convex N-gon inscribed in the unit circle. We form the new inscribed convex N-gon

$$\Psi(P) = (c_0, c_2, c_4, \dots),$$

where $c_0, c_2, c_4, \dots$ are the consecutive conformal midpoints of the points of P.

In [**S4**] we proved

THEOREM 2.2. *For any $N \geq 5$ the following is true. $\{\Psi^n(P)\}$ converges to a projectively regular N-gon for every choice of P.*

The main idea of the proof was establishing the inequality

$$\prod [c_i, c_{i+2} c_{i+4} c_{i+6}] \geq \prod [a_i, a_{i+2} a_{i+4} a_{i+6}]. \tag{2.14}$$

This inequality works for all $N \geq 5$ and one has equality iff P is projectively regular. Thus, the conformal averaging map increases the product of the cross ratios of consecutive points. The result is very similar to our Theorem 1.3, and it works for all $N \geq 5$. Equation (2.14) is what motivated Theorem 1.3.

The case $N = 5$ deserves special scrutiny. It is a fact that every pentagon is projectively equivalent to one which is inscribed in the unit circle. Moreover, the conformal averaging map is also projectively natural. We have $T\Psi = \Psi T$ whenever T is a projective transformation preserving the unit circle. Since the definition of Ψ depends on the polygon being convex, Ψ induces a map on the space $\mathcal{C}_5$ of projective classes of convex pentagon. One might wonder if Ψ is related to the projective heat map H acting on $\mathcal{C}_5$. Perhaps one has $\Psi = H$.

This turns out not to be the case. The map Ψ is not a rational map on $\mathcal{C}_5$, because the definition of the conformal midpoint involves taking a square root. The convexity constraint allows us to take a canonical choice of square root, and this makes the square-root operation less conspicuous in the construction.

CHAPTER 3

A Primer on Projective Geometry

3.1. The Real Projective Plane

The projective plane can be defined relative to any field, but we will concentrate on the case when the field is $\boldsymbol{R}$, the reals.

The *real projective plane* is the space of lines through the origin in $\boldsymbol{R}^3$. Equivalently, it is the equivalence classes of nonzero vectors, where $V \sim rV$ for any nonzero r. The projective plane is denoted $\boldsymbol{RP}^2$. We will typically denote points in $\boldsymbol{RP}^2$ by triples $[x, y, z]$. This denotes the equivalence class of the vector (x, y, z). The projective plane is a smooth compact surface.

A *line* in $\boldsymbol{RP}^2$ is the set of points represented by the lines contained in a plane through the origin. The space of lines in $\boldsymbol{RP}^2$ is often denoted $(\boldsymbol{RP}^2)^*$. We will typically denote points in $(\boldsymbol{RP}^2)^*$ by coordinates $[A, B, C]$. This denotes the line corresponding to the plane $Ax + By + Cz = 0$. The space $(\boldsymbol{RP}^2)^*$ is often called the *dual projective plane*.

The projective plane has the beautiful property that any two distinct lines intersect in a unique point, called the *meet* of the lines, and any two distinct points lie in a unique line, called the *join* of the points. We define the meet and join of A and B as (AB) in both cases. To show this notation in action, the projective midpoint of points a_{-3}, a_{-1}, a_1, a_3 is given by:

$$b_0 = ((((a_{-3}a_{-1})(a_{+1}a_{+3}))\ ((a_{-3}a_{+1})(a_{-1}a_{+3})))\ \ (a_{-1}a_{+1})). \tag{3.1}$$

See Figure 1.3.

A collection of points is called *collinear* if they all lie on the same line. A collection of lines is called *coincident* if they all contain the same point.

3.2. Affine Patches

The complement of a line in $\boldsymbol{RP}^2$ is called an *affine patch*. The most common affine patch is the subset $\boldsymbol{A}^2$ consisting of the lines $[x, y, z]$ with $z \neq 0$. There is a canonical map from $\boldsymbol{A}^2$ to $\boldsymbol{R}^2$ given by

$$[x, y, z] \to (x/z, y/z). \tag{3.2}$$

The inverse map is given by

$$(x, y) \to [x, y, 1]. \tag{3.3}$$

Under this identification, any line in $\boldsymbol{RP}^2$ which actually intersects $\boldsymbol{A}^2$ does so in an ordinary line. Using affine patches, one can do a good job of representing figures in $\boldsymbol{RP}^2$. This is exactly how Figure 1.3 works.

3.3. Projective Transformations and Dualities

Each invertible linear transformation $T: \boldsymbol{R}^3 \to \boldsymbol{R}^3$ permutes the lines through the origin and so induces a smooth diffeomorphism from $\boldsymbol{RP}^2$ to itself. Since these linear transformations also permute the planes through the origin, projective transformations permute the lines in $\boldsymbol{RP}^2$ and thus induce smooth self-diffeomorphisms of $(\boldsymbol{RP}^2)^*$. There is a nice converse to the statements above: Any homeomorphism of $\boldsymbol{RP}^2$ which maps lines to lines is a projective transformation. The proof is a fun exercise.

The group of projective transformations is denoted $PGL_3(\boldsymbol{R})$. It is the group invertible linear transformations modulo scaling. $PGL_3(\boldsymbol{R})$ is an 8-dimensional Lie group. This group acts simply transitively on the set of general position quadruples. That is, there is a unique projective transformation carrying any general position quadruple of points to any other general position quadruple of points.

Each nondegenerate quadratic form Q on $\boldsymbol{R}^3$ defines a canonical map between $\boldsymbol{RP}^2$ and $(\boldsymbol{RP}^2)^*$. The point $p \in \boldsymbol{RP}^2$, corresponds to the 0-set of the linear functional $W \to Q(V_p, W)$. Here V_p is any vector representing p. The most familiar case is when Q is the dot product. In this case the obvious map from a 1-dimensional subspace to its perpendicular complement induces the map from $\boldsymbol{RP}^2$ to $(\boldsymbol{RP}^2)^*$. Algebraically, the point $[x, y, z]$ simply corresponds to the line $[x, y, z]$ in this case. In general, these maps are called *polarities*.

A *duality* is a map of the form $\Delta = T \circ \Pi$, where Π is a polarity and T is a projective transformation. Dualities are diffeomorphisms from $\boldsymbol{RP}^2$ to $(\boldsymbol{RP}^2)^*$ which have the property of preserving incidence relations. That is, a, b, c are three collinear points in $\boldsymbol{RP}^2$ if and only if $\Delta(a), \Delta(b), \Delta(c)$ are three coincident lines. Conversely, any homeomorphism from $\boldsymbol{RP}^2$ to $(\boldsymbol{RP}^2)^*$ with the above property is a duality. Each duality automatically induces a map from $(\boldsymbol{RP}^2)^*$ to $\boldsymbol{RP}^2$: The point $\Delta(L)$ is defined to be the point which contains all the lines $\Delta(p)$ with $p \in L$. This dual map coincides with Δ^{-1} when Δ is a polarity, but in general it does not.

A *flag* is a pair (p, L), where p is a point and L is a line and $p \in L$. The *flag space* is the set all flags. The flag space is a smooth compact 3-manifold. It fibers over $\boldsymbol{RP}^2$ and $(\boldsymbol{RP}^2)^*$ in the obvious way: you either forget the line or you forget the point. Both projective transformations and dualities act as smooth diffeomorphisms of the flag space. A projective transformation T maps (p, L) to $(T(p), T(L))$ and a duality Δ maps (p, L) to $(\Delta(L), \Delta(p))$.

3.4. The Cross Ratio

The *cross ratio* of 4 real numbers $a, b, c, d \in \boldsymbol{R}$ is defined as

$$[a, b, c, d] = \frac{(a-b)(c-d)}{(a-c)(b-d)}. \tag{3.4}$$

This formula is the special case of an invariant of 4 collinear points in $\boldsymbol{RP}^2$, also called the cross ratio. Given a, b, c, d we define $[a, b, c, d] = \chi$, where

$$(\chi, \chi, \chi) = \frac{(a \times b)(c \times d)}{(a \times c)(b \times d)}, \tag{3.5}$$

equation (3.5) requires some interpretation. The quantities a, b, c, d now are vectors representing 4 collinear points in $\boldsymbol{RP}^2$. The expression $(a \times c)(b \times d)$ is the coordinatewise product of two vectors. That is, if $a \times b = (v_1, v_2, v_3)$ and

$c \times d = (w_1, w_2, w_3)$, then the product is $(v_1 w_1, v_2 w_2, v_3 w_3)$. The same goes for the denominator in equation (3.5). Finally, the ratio is the coordinatewise quotient. This quotient turns out to have all coordinates equal. If we include $\boldsymbol{R}$ as a subset of $\boldsymbol{A}^2$ in a natural way, then the two definitions of the cross ratio coincide. Moreover, $[a, b, c, d] = [T(a), T(b), T(c), T(d)]$ for any projective transformation T.

If A, B, C, D are 4 coincident lines, represented by vectors as discussed above, then $[A, B, C, D]$ can be defined using equation (3.5). This definition coincides with the cross ratio of the slopes of the lines, assuming that they all intersect the affine patch $\boldsymbol{A}^2$.

Figure 3.1 illustrates the fundamental connection between cross ratios of points and the cross ratios of lines. The cross ratio of any 4-tuple of coincident lines equals the cross ratio of the 4 points (taken in the same order) obtained by intersecting this 4-tuple with any auxiliary line.

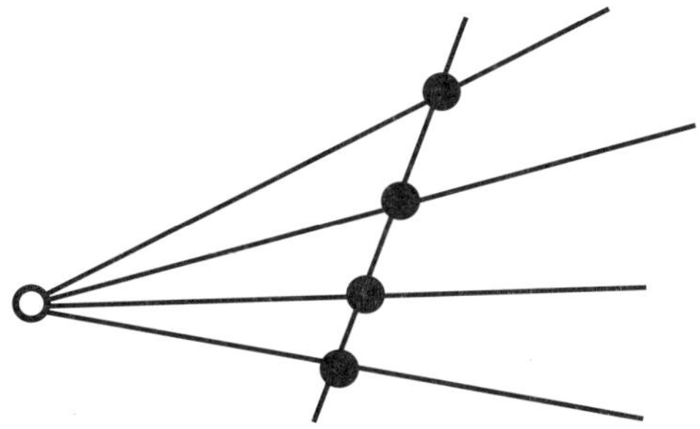

FIGURE 3.1. Cross ratios of points and lines

For later reference, we call this basic connection the *Cross Ratio Principle.*

3.5. The Hilbert Metric

A subset $S \subset \boldsymbol{RP}^2$ is called *compact convex* if there is some projective transformation T such that $T(S) \subset \boldsymbol{A}^2$ and $T(S)$ is a compact convex subset of $\boldsymbol{A}^2$ in the ordinary sense. Given a compact convex set S, there is a canonical metric on the interior S^o of S, called the *Hilbert metric*. Given points $b, c \subset S^o$, we define

$$d(b, c) = -\log \chi(a, b, c, d), \tag{3.6}$$

where a and d are the points where (bc) intersects ∂S, as shown in Figure 3.2.

When S is the unit disk, the Hilbert metric on the interior of S coincides with the Klein-Beltrami metric. This is the familiar model of the hyperbolic disk in which the geodesics are Euclidean line segments. So, one could view the Hilbert metric as a generalization of the hyperbolic metric. The following result justifies the terminology.

LEMMA 3.1. *The Hilbert metric really is a metric.*

Proof: If is not hard to see from the definition of the cross ratio that $d(b, c) \geq 0$ with equality if and only if $b = c$. Also, symmetries of the cross ratio imply that $d(b, c) = d(c, b)$. The difficult part is establishing the triangle inequality.

Suppose we want to show that $d(b, e) + d(c, e) \geq d(b, c)$. Referring to Figure 3.3, we can replace the lightly shaded region S by the darkly shaded quadrilateral

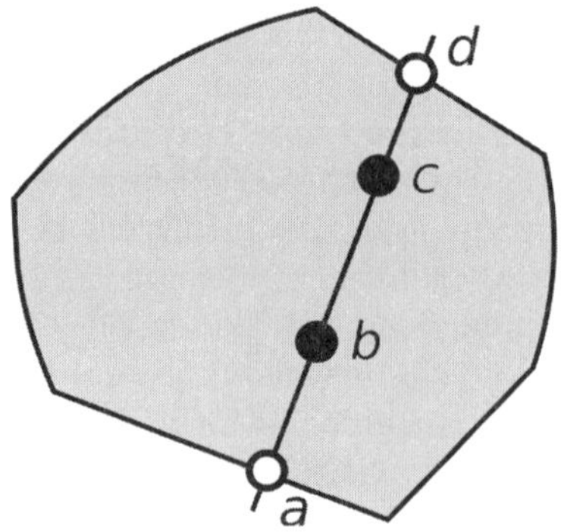

FIGURE 3.2. Construction for the Hilbert Metric

Q. In making this switch, the distances $d(b,e)$ and $d(c,e)$ do not change but, as one can easily check from the formula, $d(b,c)$ increases. So, it suffices to prove our result in the quadrilateral Q and for the points arranged as shown in Figure 3.3.

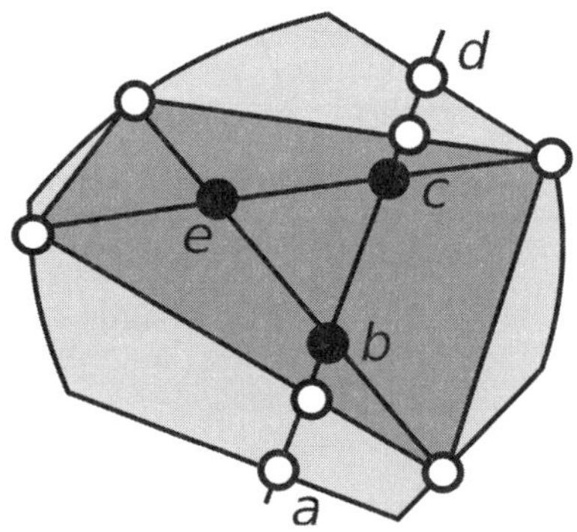

FIGURE 3.3. Switching the domain

Since every two quadrilaterals are projectively equivalent, it suffices to consider the case when $Q = [-1,1]^2$, as shown in Figure 3.4.

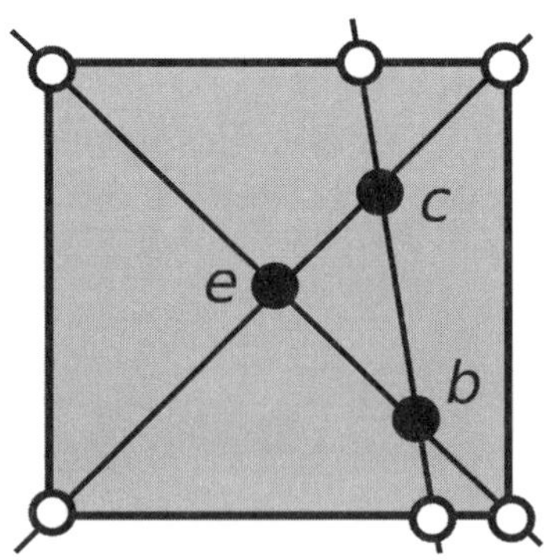

FIGURE 3.4. The normalized picture

In this case, we have

$$b = (s, -s, 1), \qquad c = (t, t, 1), \qquad e = (0, 0, 1), \quad s, t \in (0, 1).$$

We compute that

$$\exp\ d(b, e) = \frac{1-s}{1+s}, \qquad \exp\ d(c, e) = \frac{1-t}{1+t}, \qquad \exp\ d(b, d) = \frac{(1-s)(1-t)}{(1+s)(1+t)}.$$

Hence $d(b, c) = d(b, e) + d(c, e)$ in this case. $\square$

3.6. Projective Invariants of Polygons

Let P be a polygon. Figure 3.5 shows how to assign a number

$$\chi(p) = [A, B, C, D] \tag{3.7}$$

to the vertex p of P. We call these quantities the *vertex invariants* of the polygon.

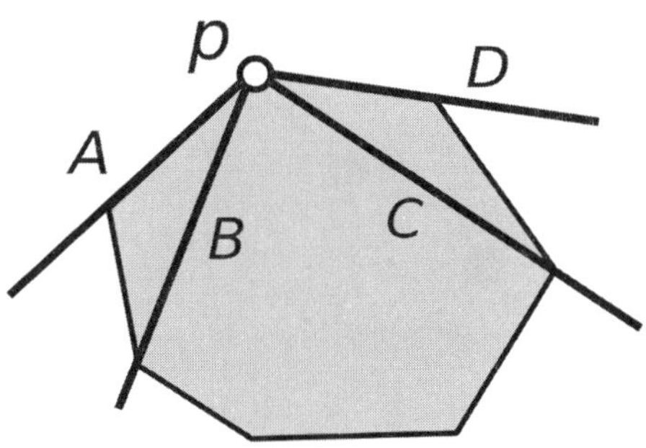

FIGURE 3.5. Definition of the vertex invariants

When P is a pentagon, two consecutive vertex invariants determine P. This is most easily seen by normalizing P so p_1, p_2, p_3, p_4 are vertices of a square and considering the two cross ratios are $\chi(p_2)$ and $\chi(p_3)$. Fixing the value of each of these cross ratios confines p_5 to a line and the two lines are distinct. So, p_5 must be the intersection of the two lines.

Figure 3.6 shows how to assign a number

$$\chi(L) = [a, b, c, d] \tag{3.8}$$

to each edge L of a polygon P. We call these quantities the *edge invariants*.

Two consecutive edge invariants determine a pentagon up to projective equivalence. Indeed, one can see from the Cross Ratio Principle that, for pentagons, $\chi(p) = \chi(L)$ whenever L is the edge opposite the point p.

On a polygon P, a *flag* is a pair (p, L) where p is a vertex and L is an edge of P incident to v. We indicate the flag (p, L) with an auxiliary white point placed on the edge L two-thirds of the way towards v. We associate the number

$$\chi(p, L) = [a, b, c, d]. \tag{3.9}$$

where a, b, c, d are as in Figure 3.7. We call these the *flag invariants* of the polygon.

Remarks:

(i) Some readers might look at Figure 3.7 and wonder if we have truly placed the auxiliary white point in the right place. Perhaps it should be 2/3 of the way to towards c rather than 2/3 of the way towards p. Let me justify the placement of

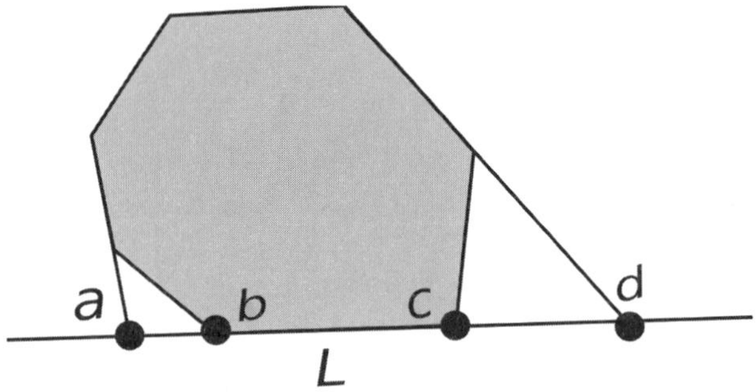

FIGURE 3.6. Definition of the edge invariants

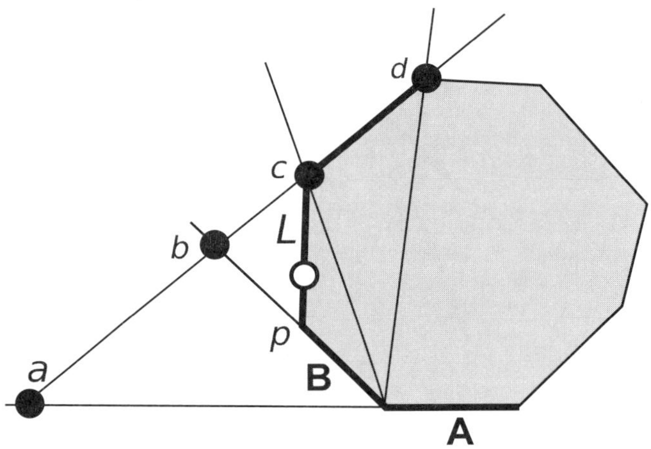

FIGURE 3.7. Invariant of a flag

the white point – i.e., the choice of flag (p, L). One could say that L is the line that is just clockwise from lines A, B and p is the vertex just counterclockwise from points c, d.
(ii) Our construction involves the edges A, B of the polygon, and the vertices c, d. In this way, the lines and vertices of the polygon are on an equal footing in the construction. We have to break the symmetry (between lines and vertices) when we take the cross ratio $[a, b, c, d]$ of points, but by the Cross Ratio Principle we could also define $\chi(p, L)$ in terms of the cross ratio of the (slopes of the) 4 lines connecting $A \cap B$ to a, b, c, d. Thus, points and lines are on a truly equal footing in the construction. We will take this up in the next section.
(iii) We have drawn these invariants for convex polygons, but these invariants make sense for any polygon in which the points are in sufficiently general position. Indeed, these invariants make sense over most fields.

In general, we list out the flag invariants of a polygon as follows. We can think of a polygon as a cyclically ordered list $v_0, L_1, v_2, L_3, \ldots$ where $v_0, v_2, v_4, \ldots$ are the vertices and $L_1, L_3, L_5, \ldots$ are the (lines extending the) edges between them – i.e. L_{2k+1} is incident to both v_{2k} and v_{2k+2}. The flag invariants are then $\chi(v_0, L_1)$,

$\chi(v_2, L_1)$, $\chi(v_2, L_3)$,... The invariants come in the same order as the auxiliary white dots representing the flags.

Here are the fundamental relations between these invariants.

LEMMA 3.2. *Suppose that p_1 and p_2 are two consecutive vertices which are both incident to the edge L. Suppose also that L_1 and L_2 are two consecutive edges which are incident to the vertex p. Then*

$$\chi(p, L_1)\chi(p, L_2) = \chi(p), \qquad \chi(p_1, L)\chi(p_2, L) = \chi(L). \tag{3.10}$$

Proof: The first relation only involves 5 consecutive points. So, if the first relation holds for all pentagons, it holds for all polygons. For pentagons, we just have to check the case when $p = p_5$ and

$$p_1 = (0,1), \qquad p_2 = (0,0), \qquad p_3 = (1,0), \qquad p_4 = (1,1), \qquad p_5 = (x,y). \tag{3.11}$$

The flag invariants associated to the flags incident to p_5 are

$$\frac{x-1}{x-y}, \qquad \frac{x}{x+y-1}. \tag{3.12}$$

The vertex invariant is

$$\frac{(x-1)x}{(x-y)(x+y-1)}, \tag{3.13}$$

namely the product of the two flag invariants. This establishes the first relation. The second relation follows from the first relation and from projective duality considerations explained in the next section. (A direct calculation would also work.) □

3.7. Duality and Relabeling

The flag invariants are obviously invariant under projective transformations. Moreover, they are also invariant under dualities, once a proper interpretation is given. First of all, we make the labeling convention that the flag invariants $x_1, x_2, \ldots$ of an N-gon P always have the property that x_1 and x_2 correspond to flags which share a common edge. The opposite convention would stipulate that the first two invariants correspond to flags which share a common vertex.

Given a duality Δ, we let $\Delta^*(P)$ denote the polygon whose edges are the edges of $\Delta(P)$. A better way to understand the action of Δ^* is to think of P as a list of flags. When P is an N-gon, the flag perspective on P realizes P as a $2N$-gon in the flag space. Denote this $2N$-gon by $P^\#$. Let $\Delta^\#$ denote the action of Δ on the flag space, as discussed at the end of §3.3. We have

$$(\Delta^*(P))^\# = \Delta^\#(P^\#). \tag{3.14}$$

Inspecting our construction of our flag invariants given in Figure 3.7, we we see that P and $\Delta^*(P)$ have the same flag invariants. Here is the proof. Label the vertices of Figure 3.7 clockwise so that $p = p_2$. Label the edges of Figure 3.7 counter-clockwise so that $L = L_2$. Then

$$\chi(p, L) = [((p_0p_1)(p_3p_4)), ((p_1p_2)(p_3p_4)), p_3, p_4] =$$
$$[((L_0L_1)(L_3L_4)), ((L_1L_2)(L_3L_4)), L_3, L_4].$$

The construction remains unchanged when the roles of points and lines are swapped.

Suppose that the flag invariants of P are $(x_1, \ldots, x_{2N})$. If we cyclically relabel the vertices of P the new flag invariants could be $(x_{2k+1}, x_{2k+2}, \ldots)$ for some integer k. That is, cyclically relabeling the vertices corresponds to shifting the indices by an even integer.

Which polygon has the flag invariants $(x_2, x_3, \ldots)$? Certainly it is not the polygon P with a different labeling of the vertices. We have already accounted for what happens when we cyclically relabel the vertices of P. The answer is that these are the invariants for $\Delta^*(P)$ relative to any duality, provided that a suitable labeling of the vertices of $\Delta^*(P)$ has been made.

3.8. The Gauss Group

Let $x_1, x_2, \ldots$ be the flag invariants associated to a pentagon. A direct calculation reveals that

$$x_{k+2} = \frac{1 - x_k}{1 - x_k x_{k+1}}. \tag{3.15}$$

Equivalently,

$$(x_{k+1}, x_{k+2}) = G(x_k, x_{k+1}), \qquad G(x, y) = \left(y, \frac{1 - x}{1 - xy}\right). \tag{3.16}$$

The map G is a famous map. It is often called the *Gauss Recurrence*. It arose in Gauss's study of spherical pentagons. The topic goes under the heading of Gauss's *pentagramma mirificum*. Beautifully, G has order 5. This means that the flag invariants for a pentagon repeat after 5 steps. They are

$$x_1, x_2, x_3, x_4, x_5, x_1, x_2, x_3, x_4, x_5.$$

We define the *Gauss group* to be the group generated by the elements G and the order 2 element

$$R(x, y) = (y, x). \tag{3.17}$$

The Gauss group has order 10. Geometrically, the Gauss group records the action of the dihedral relabeling group on the space of $\mathcal{P}_5$ of labeled pentagons modulo projective equivalence. However, there is a subtlety here. Cyclically shifting the vertices corresponds to an even power of the Gauss recurrence, but the fact that the Gauss recurrence has order 5 means that odd powers of the Gauss recurrence are also even powers and hence are realized by cyclic relabellings as well.

There is a map from $\mathcal{P}_5$ into $\boldsymbol{R}^2$ given by

$$f(P) = (\chi(p_1, L), \chi(p_2, L)). \tag{3.18}$$

That is, we just take two consecutive flag invariants. f conjugates the group of dihedral relabelings to the Gauss group.

The fact that G^5 is the identity also has a geometric explanation. We have already mentioned that for pentagons $\chi(p) = \chi(L)$ if the vertex v is opposite the edge L. This translates into the statement that there is a projective duality which carries the vertices of P to the lines of P. That is, $P = \Delta^*(P)$. Given the action of Δ^* on the flag invariants, discussed in the previous section, we see that the list of flag invariants of P must repeat after an odd number k of steps. The only possibility is that $k = 5$.

CHAPTER 4

Elementary Algebraic Geometry

In this chapter we present some results and definitions from elementary algebraic geometry. Some of the results we will quote without proof and sometimes we will give proofs. See W. Fulton's book [**F**] for a much more thorough treatment. See also K. Kendig's book [**K**].

4.1. Measure Zero Sets

A subset $S \subset \boldsymbol{R}^n$ has *measure zero* if, for any $\epsilon > 0$, the set S is contained in a countable union of cubes such that the sum of the volumes of the cubes is less than ϵ. On the other extreme, S has *full measure* if $\boldsymbol{R}^n - S$ has measure zero. Countable unions of measure zero sets have measure zero, and (hence) countable intersections of full measure sets have full measure.

If M is a smooth manifold, a subset $S \subset M$, which happens to be contained in a coordinate chart (U, ϕ) of M, has measure 0 provided that $\phi(S) \subset \boldsymbol{R}^n$ has measure zero. A general subset $S \subset M$ has measure 0 if S is the countable union of measure zero sets contained in coordinate charts. A full measure subset of M is one whose complement has zero measure.

We say that *almost every point has property P* provided that property P fails only on a set of measure zero. For instance, a nonconstant polynomial on $\boldsymbol{R}^n$ is nonzero almost everywhere.

4.2. Rational Maps

A *rational function* on $\boldsymbol{R}^n$ is a function of the form $f = P/Q$ where both P and Q are polynomials. Such maps are defined on the set where Q is nonzero. That is, they are defined except on a set of measure 0. A *rational map* from $\boldsymbol{R}^n$ to $\boldsymbol{R}^n$ is a map of the form $f = (f_1, \ldots, f_n)$ where each f_j is a rational function. A rational map is defined almost everywhere.

A rational map $f : \boldsymbol{R}^n \to \boldsymbol{R}^n$ is called *birational* if there is another rational map $f : \boldsymbol{R}^n \to \boldsymbol{R}^n$ such that fg and gf are the identity map wherever both are defined. Often we write $g = f^{-1}$. There is an open and full measure subset where both maps are defined.

The composition of rational maps is again a rational map. Hence, the set of all rational maps on $\boldsymbol{R}^n$ forms a semigroup and the set of birational maps on $\boldsymbol{R}^n$ forms a group. This group is called the *Cremona group*. The Gauss group considered at the end of the last chapter is an order 10 subgroup of the Cremona group of $\boldsymbol{R}^2$.

Given a rational map f, and $x \in \boldsymbol{R}^n$, we define $f^n(x) = f \circ \cdots \circ f(x)$, provided that all the maps are well-defined on the relevant points. The *forward orbit* of x is $\{f^n(x)\}$ provided that this is well defined. When f is birational, we define

$f^{-n} = (f^{-1})^n$ and we define the *orbit* of x to be the bi-infinite sequence $\{f^n(x)\}$. Almost every point in $\boldsymbol{R}^n$ has a well-defined orbit.

4.3. Homogeneous Polynomials

A *homogeneous polynomial* in 3 variables is any finite sum

$$P(x, y, z) = \sum c_{i,j,k} x^i y^j z^k, \tag{4.1}$$

where $c_{i,j,k} \in \boldsymbol{C}$ and the total sum $i + j + k = D$ is independent of the summand. D is called the *degree* of P. Though we view P as a map from $\boldsymbol{C}^3$ to $\boldsymbol{C}$, it makes sense to talk about the *zero set* of P as a subset of the complex projective plane $\boldsymbol{CP}^2$ where P vanishes. That is,

$$V(P) = \{[v] \in \boldsymbol{CP}^2 | \; P(v) = 0\}. \tag{4.2}$$

Here $\boldsymbol{CP}^2$ is the space of complex lines through the origin in $\boldsymbol{C}^3$. The definition of $V(P)$ makes sense because $P(v) = 0$ if and only if $P(\lambda v) = 0$ for any nonzero $v \in \boldsymbol{C}^3$ and any nonzero $\lambda \in \boldsymbol{C}$. The subset $V(P) \subset \boldsymbol{CP}^2$ is a special case of what is called a *projective variety*.

Each homogeneous polynomial on $\boldsymbol{C}^3$ gives rise to an ordinary polynomial on $\boldsymbol{C}^2$ just by setting $z = 1$. Formally, this amounts to restricting the polynomial to the affine patch $\{z \neq 0\}$ and then conjugating by the canonical map to $\boldsymbol{C}^2$. This 2-variable polynomial is called the *dehomogenization* of the homogeneous polynomial.

The process can be reversed. Given a 2 variable polynomial in x, y, we can simply pad it with powers of z to make it homogeneous. One example should explain the whole construction. Suppose that $f(x, y) = x^2 + 3xy^2 + 2$. Then the homogeneous polynomial is $F(x, y, z) = x^2 z + 3xy^2 + 2z^3$. The polynomial F is called the *homogenization* of f.

There are two other equally natural affine patches, namely the set $\{x \neq 0\}$ and the set $\{y \neq 0\}$. The homogenenization and dehomogenization process works the same for these affine patches. Note that $\boldsymbol{CP}^2$ is covered by these three affine patches. So, if we want to understand the zero set of a homogeneous polynomial in $\boldsymbol{CP}^2$, it suffices to study the inhomogeneous polynomials obtained by setting each of the coordinates equal to 1.

4.4. Bezout's Theorem

Let $\boldsymbol{C}[x, y]$ denote the set of polynomials in 2 variables with complex coefficients. Given $f \in \boldsymbol{C}[x, y]$ we let $V(f)$ denote the solution set of the equation $f = 0$. The set $V(f)$ is called a *plane algebraic curve*. Let $V(f_1, f_2)$ denote the solution set of the equations $f_1 = f_2 = 0$. When f_1 and f_2 have no common factors, $V(f_1, f_2)$ is a finite set of points. We explain how to properly count the number of points in $V(f_1, f_2)$. The quantity of interest to us is often denoted $I_p(f_1, f_2)$, and it is called the *intersection number* of $V(f_1)$ and $V(f_2)$ at p.

An algebraic definition of the intersection number $I_p(f_1, f_2)$ is given in [**F**]. For the sake of exposition, we give a more analytic definition which doesn't require a build-up of algebraic geometry. The equivalence of the analytic and algebraic definitions can be extracted from [**AGV**, §5].

We define a *δ-perturbation* of a polynomial f_1 to be a polynomial g_1 with the property that $\deg(f_1) = \deg(g_1)$ and all the coefficients of $f_1 - g_1$ have norm less than δ. We define

$$I_p(f_1, f_2) = \min_{\epsilon>0} \left(\limsup_{\delta\to 0} N(g_1, g_2, \epsilon) \right). \tag{4.3}$$

Here $N(g_1, g_2, \epsilon)$ denotes the number of solutions to $g_1 = g_2 = 0$ within ϵ of p, when g_1 and g_2 are δ-perturbations of f_1 and f_2. Intuitively, we perturb f_1 and f_2 in a generic way and count the number of points in $V(g_1, g_2)$ near p.

When it comes time for us to compute some intersection numbers, we will not use the definition above. Rather we just one of the properties of the intersection number listed on p 37 of [**F**]:

$$I_p(f_1, f_2) \geq m_p(f_1)m_p(f_2). \tag{4.4}$$

The quantity $m_p(f_j)$ is known as the *multiplicity* of f_j at p. All we need to know about this quantity is as follows: If f_j vanishes to kth order at p then $m_p(f_j) \geq k$. In the application the point of interest will always be $(0,0)$, and to show that $m_{(0,0)}(f_j) \geq k$ we just have to show that

$$f_j(x, y) = g_k(x, y) + \text{higher order terms.} \tag{4.5}$$

Here g_k is a homogeneous polynomial in two variables of degree k.

If f_1 and f_2 are homogeneous polynomials in 3 variables with no common factors and $p \in \boldsymbol{CP}^2$ is a solution to the simultaneous equation $f_1 = f_2 = 0$, we define $I_p(f_1, f_2) = I_{p^*}(f_1^*, f_2^*)$ where p^* is the image of p in $\boldsymbol{C}^2$ under the identification of $\boldsymbol{C}^2$ with an affine patch containing p. Here f_1^* and f_2^* are the corresponding inhomogeneous polynomials. This definition is independent of the chosen affine patch. As discussed, we always use one of the three affine patches mentioned above.

Here is a special case of Bezout's Theorem.

THEOREM 4.1 (Bezout). *Let f_1 and f_2 be two homogeneous polynomials in $\boldsymbol{C}[x, y, z]$ with no common factors. Then*

$$\sum_{p\in V(f_1,f_2)} I_p(f_1, f_2) = \deg(f_1)\deg(f_2). \tag{4.6}$$

We emphasize that the count of the intersections takes place in all of $\boldsymbol{CP}^2$. Practically any book on algebraic geometry has a proof of Bezout's Theorem.

4.5. The Blow-up Construction

The blow-up construction starts with a smooth manifold M and some point $p \in M$ and returns a new smooth manifold M_p. The reader should know that the construction also makes sense in the category of algebraic manifolds. An *algebraic manifold* is a smooth manifold equipped with an atlas of coordinate charts whose overlap functions are birational maps. For instance, the space $\boldsymbol{RP}^n$ of lines through the origin in $\boldsymbol{R}^{n+1}$ is naturally an algebraic manifold. If M is an algebraic manifold, then M_p naturally inherits the structure of an algebraic manifold. We will just use the term *manifold* to refer to manifolds of either type.

We first discuss $\boldsymbol{R}^n_0$, the blowup of $\boldsymbol{R}^n$ at the origin. We denote the origin by 0, through really the point is $(0, \ldots, 0)$. As a set, $\boldsymbol{R}^n_0$ is the set of flags (p, L) where $p \in \boldsymbol{R}^n$ and $L \in \boldsymbol{RP}^{n-1}$ and $p \in L$. This set is naturally a submanifold of the product $\boldsymbol{R}^n \times \boldsymbol{RP}^{n-1}$.

There is a canonical map $\pi : \boldsymbol{R}^n_0 \to \boldsymbol{R}^n$ given by

$$\pi(p, L) = p. \tag{4.7}$$

We call π the *blow-down map*. Since each nonzero point in $\boldsymbol{R}^n$ determines a unique line through the origin, π is a birational diffeomorphism from $\boldsymbol{R}^n_0 - \pi^{-1}(0)$ to $\boldsymbol{R}^n - \{0\}$. The inverse image $\pi^{-1}(0)$ is a copy of $\boldsymbol{RP}^{n-1}$. It is called the *exceptional fiber*. If $B \subset \boldsymbol{R}^n$ is some open set which contains the origin, we let $B_0 = \pi^{-1}(B)$. We think of B_0 as the blowup of B at 0.

Now consider the general case of a manifold M with $p \in M$. The easiest way to understand the blowup M_p is to choose a coordinate chart on M such that p is (identified with) the origin in $\boldsymbol{R}^n$ and some open ball $B \subset \boldsymbol{R}^n$ about the origin is (identified with) an open neighborhood of p in M. We get M_p by cutting out B and pasting back in B_0. Formally, M_p is the quotient space

$$\Big((M - B) \coprod B_0\Big)/\sim, \tag{4.8}$$

where $q_1 \sim q_2$ if and only if $q_1 \in B - 0$ and $q_2 \in B_0 - \pi^{-1}(0)$ and $\pi(q_2) = q_1$. Here $\coprod$ denotes the disjoint union.

A bit of reflection reveals that the new space can be naturally be made into a manifold again, and the isomorphism class of the manifold (in whatever category) does not depend on the choice of B. Notice also that the above construction would work about the same way if we chose a coordinate chart in which a neighborhood of $p \in M$ was identified with an open ball B' about some other point in $\boldsymbol{R}^n$. We would blow up by replacing B' with a suitably translated copy $(B_0)'$ of B_0.

Later on, we will blow up $(\boldsymbol{R} \cup \infty)^2$ at 3 points. Note that $\boldsymbol{R}^2$ is naturally a subset of $(\boldsymbol{R} \cup \infty)^2$. If the points all lie in $\boldsymbol{R}^2$ we can perform the blow-up operation as above, and simultaneously, using pairwise disjoint disks centered at these points. The first time we do this triple blowup, the points we consider, namely $(1,1)$ and $(0,\infty)$ and $(\infty,0)$, do not all belong to $\boldsymbol{R}^2$. However, we could move these points into $\boldsymbol{R}^2$ using a suitable birational change of coordinates – e.g. the map $\boldsymbol{B}$ from equation (1.2). Indeed, the case we really care about is when we blow up $(\boldsymbol{R} \cup \infty)^2$ at the points $\boldsymbol{B}(1,1)$, $\boldsymbol{B}(0,\infty)$ and $\boldsymbol{B}(\infty,0)$, all of which do lie in $\boldsymbol{R}^2$. The first blow-up construction is just a stepping stone to understanding the second one.

Here we explain one use of the blow-up construction. We stick to the 2 dimensional case and we imagine that we have blown up $\boldsymbol{R}^2$ at some point $p \in \boldsymbol{R}^2$. We say that a rational function $f = P/Q$ has a *simple blowup* at p if $P(p) = 0$ and $Q(p) = 0$ and the gradients $\nabla P(p)$ and $\nabla Q(p)$ are linearly independent.

LEMMA 4.2. *Suppose that f has a simple blowup at p. There is an open neighborhood U of $\pi^{-1}(p)$ in $\boldsymbol{R}^2_p$ and a smooth map $f_p : U \to \boldsymbol{R} \cup \infty$ so that $\pi \circ f_p = f$ on $\pi(U) - p$.*

Proof: After making an affine change of coordinates, we can assume that $p = (0,0)$ and

$$f(x,y) = \frac{y + \sum_{i+j\ge 2} a_{ij}x^iy^j}{x + \sum_{i+j\ge 2} b_{ij}x^iy^j}. \tag{4.9}$$

The blowup $\boldsymbol{R}^2_0$ is the space of flags (q, ℓ) where $q \in \boldsymbol{R}^2$ and ℓ is a line through $(0,0)$ and q. We write $\ell = \{(tu, tv)|\ t \in \boldsymbol{R}\}$ and denote the point $((0,0), \ell)$ as $[u : v]$. We have $\pi(q,\ell) = q$. We define $f_0(q,\ell) = f(q)$ when $q \neq (0,0)$ and $f_0([u:v]) = u/v$. Evidently, $f = \pi \circ f_0$ whenever f is defined.

Since π is a diffeomorphism away from $\pi^{-1}(0)$, the map f_0 is smooth away from $\pi^{-1}(0)$. We just have to see that f_0 is smooth at an arbitrary point $[u:v]$. Replacing f with $1/f$ if necessary, we can without loss of generality consider the case when $|v| < |u|$. We introduce smooth local coordinates in a neighborhood of $[u:v] \in \boldsymbol{R}_0^2$ which have the form (x,t). When $x \neq 0$ the point (x,t) corresponds to (x,tx) and when $x = 0$ the point (x,t) corresponds to $[1:t]$. Since we want to work in an open neighborhood of $[u:v]$, we can take, say, $|t| \leq 2$.

In our new coordinates, we have

$$f_0(x,t) = \frac{t + \sum_{i+j\geq 2} a_{ij} t^j x^{i+j-1}}{1 + \sum_{i+j\geq 2} b_{ij} x^{i+j-1} t^j} = t + \sum_{k\geq 1} P_k(t) x^k. \tag{4.10}$$

Here $P_k(t)$ is a polynomial in t. The degree and maximum size of a coefficient in P_k grows at most exponentially. So, for $|x|$ sufficiently small, the equation above defines a convergent power series and hence a smooth map. □

CHAPTER 5

The Pentagram Map

5.1. The Pentagram Configuration Theorem

The Pentagram Configuration Theorem says that a pentagon P and its pentagram P' are projectively equivalent. I learned the proof from John Conway in 1988. See also [**Mot**] or [**S1**]. Figure 5.1 shows the proof. By two applications of the Cross Ratio Principle, the vertex invariant of the outer pentagon P at the highlighted black vertex is the same as the vertex invariant of the inner pentagon P' at the highlighted black vertex. In both cases, the invariant is the cross ratio of the 4 white points. Hence P and P' have the same vertex invariants. Hence, they are projectively equivalent.

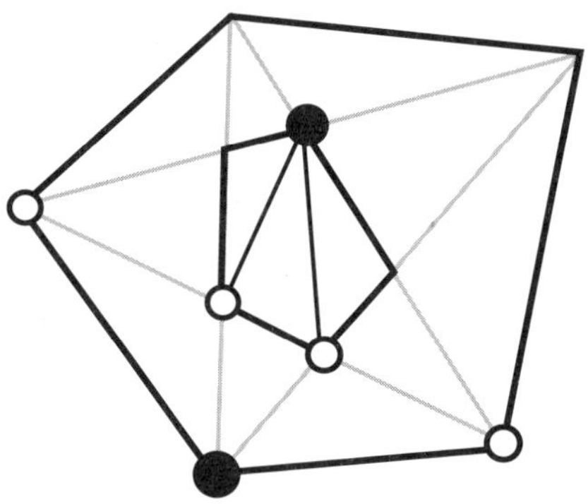

FIGURE 5.1. Two vertex invariants coincide

There is another configuration theorem like this. The two hexagons P and P'' are always projectively equivalent when they are labeled as in Figure 5.2. I'll leave this to the interested reader. Actually, I don't know a geometric proof. See [**ST**] for some other related configuration theorems.

5.2. The Pentagram Map in Coordinates

By now there are many papers on the pentagram map. See, for instance, [**S1**], [**S2**], [**S3**], [**OST1**], [**OST2**], [**Sol1**], [**KS**], [**Gli**], [**GSTV**], [**MB1**], [**MB2**].

Let T denote the pentagram map acting on the space $\mathcal{P}_N$ of projective equivalence classes of labeled N-gons. As for the projective heat map there is no canonical way to make T act on $\mathcal{P}_N$. One must break the symmetry to get a labeling scheme. Whatever the labeling scheme, T is not periodic for $N \geq 7$.

Figure 5.3 shows a canonical labeling scheme for T^2. Note that this is different from the labeling scheme in Figure 5.2. The labeling scheme in Figure 5.2 is designed

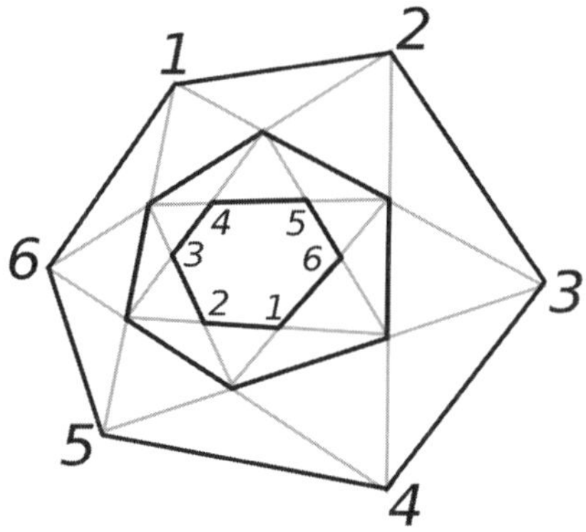

FIGURE 5.2. The pentagram map acting on hexagons

specially for $\mathcal{P}_6$. Were we to use the labeling scheme from Figure 5.3 for $\mathcal{P}_6$, the map T^2 would have order 2.

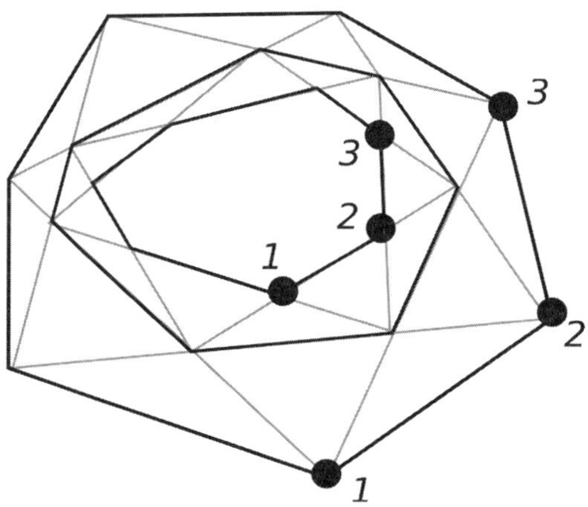

FIGURE 5.3. Canonical labeling scheme for T^2

Our discussion now refers to the invariants introduced in §3.6. Let $x_1, x_2, \ldots$ be the flag coordinates for an N-gon P. In [**S3**] we show that

$$T^2 = \alpha_1 \circ \alpha_2. \tag{5.1}$$

Here $\alpha_1(x_1, \ldots, x_{2n}) = (x'_1, \ldots, x'_{2n})$ and $\alpha_2(x_1, \ldots, x_{2n}) = (x''_1, \ldots, x''_{2n})$ where

$$x'_{2k-1} = x_{2k} \frac{1 - x_{2k+1}x_{2k+2}}{1 - x_{2k-3}x_{2k-2}}; \qquad x'_{2k} = x_{2k-1} \frac{1 - x_{2k-3}x_{2k-2}}{1 - x_{2k+1}x_{2k+2}};$$

$$x''_{2k+1} = x_{2k} \frac{1 - x_{2k-2}x_{2k-1}}{1 - x_{2k+2}x_{2k+3}} \qquad x''_{2k} = x_{2k+1} \frac{1 - x_{2k+2}x_{2k+3}}{1 - x_{2k-2}x_{2k-1}} \tag{5.2}$$

Both α_1 and α_2 are involutions. Thus, T^2 is the composition of two involutions.

Remarks:

(i) The pentagram map gives rise to a birational map of $\boldsymbol{R}^{2n}$. Not all points in $\boldsymbol{R}^{2n}$ represent the flag invariants of an n-gon. We discuss this in detail in §5.7.

(ii) Some authors (including myself) of pentagram map papers call the flag invariants defined above the *corner invariants*. I am calling these invariants the flag invariants here to distinguish them from the vertex invariants defined above. Also, since the flag invariants are associated to the flags of the polygon, and the vertex invariants are associated to the vertices, the terminology here is better.

5.3. The First Pentagram Invariant

For each $P \in \mathcal{P}_N$, let

$$f(P) = \prod_{i=1}^{2n} x_i \tag{5.3}$$

denote the product of the flag invariants associated to the N-gon P. It follows immediately from the formulas that $f(P) = f(T(P))$. That is, f is an invariant of the pentagram map. In papers on the pentagram map, f is written as O_5E_5. See e.g. [**S3**] or [**OST**], and also §5.7 below.

Each vertex is incident to two flags. As we have already seen in §3.6, each vertex invariant of P is the product of the flag invariants associated to these two incident flags. Therefore, $f(P)$ is also the square of the product of the vertex invariants. Tracing through the definitions, we see that $-\frac{1}{2}\log(f(P))$ is the perimeter of P' in the Hilbert metric on (the open region bounded by) P.

Remark: The only other thing in this chapter that is used in later in the monograph is the notion of a twisted polygon defined in §5.6. We mention this because some of the material below is rather advanced and also a bit sketchy.

THEOREM 5.1. *If P is a convex N-gon, then $\{T^n(P)\}$ shrinks to a point.*

Proof: If this is false, then $\bigcap T^n(P) = K$, some nontrivial compact convex subset. Let L be any line that intersects K in a nontrivial segment. Then $L \cap T^n(P)$ converges to the segment $L \cap K$. In particular, the endpoints of $L \cap T^{n+1}(P)$ converge to the respective endpoints of $T^n(P)$, and these endpoints are distinct. This is enough to show that the Hilbert diameter of $T^{n+1}(P)$ in $T^n(P)$ tends to ∞. But then, by the triangle inequality, the Hilbert perimeter of $T^{n+1}(P)$ in $T^n(P)$ tends to ∞, a contradiction. □

LEMMA 5.2. *The space $\mathcal{C}_N$ is diffeomorphic to $\boldsymbol{R}^{2N-8}$.*

Proof: To see this, we normalize the first 4 points by a projective transformation so that they are the vertices of a square. This leaves $2N-8$ degrees of freedom to pick the remaining points, and accounts for the dimension. The fifth point is forced to lie in a triangle $T \subset \boldsymbol{RP}^2$ bounded by certain of the lines extending the edges of the polygon in order to retain the convexity condition. Once the fifth point is placed, the 6th point must lie in a triangle. And so on. Thus, $\mathcal{C}_N$ is an open triangle bundle over $\mathcal{C}_{N-1}$. The topological part of the lemma follows from this. □

The flag invariants lie in $(0,1)$ for convex polygons, so f maps $\mathcal{C}_N$ into $(0,1)$.

LEMMA 5.3. *f is a proper function on $\mathcal{C}_N$.*

Proof: A careful proof is given in [**S1**]. Here is a sketch. It suffices to show that $f(P_n) \to 0$ on any sequence P_n which exits every compact subset of $\mathcal{C}_N$. Since all the flag invariants are in $(0,1)$ it suffices to show that at least one flag invariant tends to 0 with n. We normalize so that the first 4 points are the vertices if the unit square. Let $p_{n,k}$ denote the kth vertex of P_n. Let T_n be the triangle from Lemma 5.2, so that $p_{n,5} \in T_n$. After cyclically relabeling the points if necessary, and passing to a subsequence, we can arrange that p_n converges to a point on ∂T_n. In other words, at least one 5-tuple must be degenerating in a projective sense and we relabel to arrange that $(p_{n,1},\ldots,p_{n,5})$ is a degenerating 5-tuple. A case-by-case analysis shows that one of the vertex invariants $\chi(p_{n,4}), \chi(p_{n,4}), \chi(p_{n,5})$ tends to 0. But each vertex invariant is the product of two flag invariants. Hence, some flag invariant tends to 0 as well. □

Lemma 5.3 says that the iterates $T^n(P)$ stay more or less the same shape in a projective sense, as they shrink to a point. Now we will go more deeply into the dynamics of the pentagram map.

5.4. The Poincare Recurrence Theorem

As a prelude to proving the recurrence of the pentagram map, we give a proof of the Poincare Recurrence Theorem. Let M be some topological space and let $T : M \to M$ be a map. A point $p \in M$ is *recurrent* for T if p is an accumulation point of its own forward orbit.

THEOREM 5.4 (Poincare Recurrence). *Suppose M is a topological space with a countable basis for its topology and a finite measure which assigns positive values to all open sets. Suppose that $T : M \to M$ is an invertible and measure-preserving map of M. Then almost every point of M is recurrent for T.*

Proof: Let $\{U_j\}$ be a countable basis of open sets for the topology. Say that $p \in U_j$ is *good* with respect to U_j if $T^k(p) \in U_j$ for infinitely many $k > 0$, and otherwise *bad* with respect to U_j. We will prove that the set of points that are bad with respect to U_j has measure 0. Say that a point $p \in M$ is *bad* if it is bad with respect to some U_j that contains it. The countable union of sets of measure zero again has measure zero. Hence, the set of bad points has measure 0. Any point that is not bad is good with respect to every basis element which contains it, and hence is a recurrent point. So, almost every point is a recurrent point.

Now we will prove that the set of points bad with respect to U_j has measure 0. Set $U = U_j$. Let

$$V_m = \bigcup_{k=m}^{\infty} T^{-k}(U), \qquad V = \bigcap_{m=0}^{\infty} V_m. \tag{5.4}$$

Note that $U \subset V_0$. We have $V_0 \supset V_1 \supset V_2 \ldots$ and all these sets have finite measure. Moreover $T(V_m) = V_{m-1}$ for all m. Hence, all these sets have the same measure. But then the set of points in $V_0 - V$ has measure 0. Since U has positive measure, almost all points of U lie in V. But every point in V is good with respect to U. □

5.5. Recurrence of the Pentagram Map

Now we sketch the proof that the pentagram map is recurrent on $\mathcal{C}_N$. A complete proof is given in [**S2**].

LEMMA 5.5. *There exists a T-invariant measure on $\mathcal{C}_N$. The measure comes from a smooth volume form.*

Proof: Here is a sketch. Let $\mathcal{X}_N$ denote the space of all convex N-gons in $\boldsymbol{RP}^2$. We have $\mathcal{C}_N = \mathcal{X}_N/PGL_3(\boldsymbol{R})$. We will construct a volume form on $\mathcal{X}_N$ which is both T-invariant and $PGL_3(\boldsymbol{R})$-invariant. We can describe the tangent space to $\mathcal{X}_N$ at a point P as the set of infinitesimal variations of a polygon P. An infinitesimal variation is specified by $2N$ vectors, with 2 based at each vertex of P. Figure 5.4 shows one of $2N$ canonical variations. All the points of P stay fixed except one, which moves according to the function $t \to p + tV$. Here V is tangent to the segment AB at p and is a unit vector relative to the Hilbert metric on AB. The other canonical variations have the same kind of description.

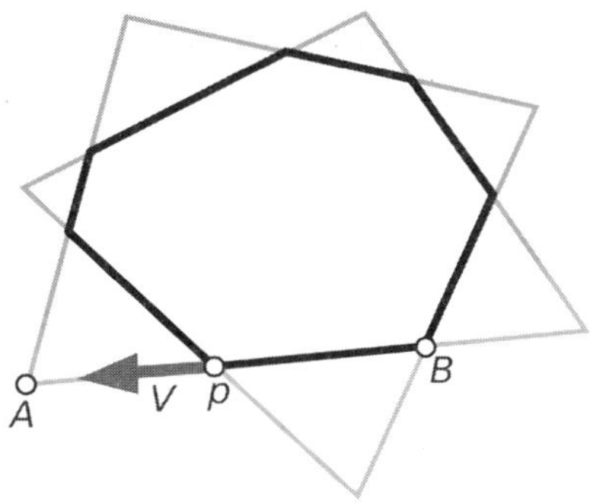

FIGURE 5.4. One of the canonical variations

When we take the $2N$ canonical variations together we get a $PGL_3(\boldsymbol{R})$ invariant framing of $\mathcal{X}_N$ – i.e. a basis of each tangent space. When we write the differential dT relative to this canonical basis (for P and for P') we compute that $\det(dT) = \pm 1$. So, we can define a $PGL_3(\boldsymbol{R})$-invariant and (up to sign) T-invariant volume form ω by requiring that ω evaluates to 1 on the canonical basis. The unsigned measure associated to ω is T-invariant. □

For almost every $\epsilon \in (0,1)$ the T-invariant set $K_\epsilon = f^{-1}[\epsilon, 1)$ is a smooth manifold with boundary. This is a consequence of Sard's Theorem. The space K_ϵ has a finite volume smooth invariant volume form. In particular, the map $f : K_\epsilon \to K_\epsilon$ is a measure preserving transformation acting on a space of finite measure. It now follows from the Poincare Recurrence Theorem that the pentagram map is recurrent on $\mathcal{C}_N$. Now we have a little more information about the shapes in the orbit $\{T^n(P)\}$. Typically one sees nearly the same shape repeat over and over.

The rest of this chapter is devoted to sketching the complete integrability of the pentagram map on $\mathcal{C}_N$ (and related spaces). The complete integrability implies a much stronger kind of recurrence.

5.6. Twisted Polygons

A *twisted N-gon* is a map $\Psi : \boldsymbol{Z} \to \boldsymbol{RP}^2$ such that

$$\Psi(k+N) = M \circ \Psi(k), \qquad \forall k \in \boldsymbol{Z}. \tag{5.5}$$

Here $M \in PGL_3(\boldsymbol{R})$ is a projective transformation called the *monodromy* of Ψ. When M is the identity, the notions of twisted and ordinary N-gons coincide. I introduced twisted polygons in [**S3**] and they arise in almost every discussion of the pentagram map.

Two twisted N-gons Ψ_1 and Ψ_2 are *equivalent* if there is a projective transformation S such that $S \circ \Psi_1 = \Psi_2$. In this case, $M_2 = SM_1S^{-1}$. Here M_j is the monodromy of Ψ_j. In other words, projectively equivalent twisted N-gons have conjugate monodromies.

Generically, any list $x_1, \dots, x_{2N}$ of real numbers coincides with $2N$ consecutive flag invariants of a twisted N-gon. (The infinite list of invariants is $2N$-periodic.) The twisted N-gon is uniquely determined by the list of numbers, up to projective equivalence. We will have more to say about this in the next chapter. Thus, we naturally identify $\boldsymbol{R}^{2N}$ with the space $\mathcal{Z}_N$ of twisted N-gons modulo projective equivalence. Some points, such as $(0, \dots, 0)$, do not correspond to geometrically realized twisted N-gons, but we are after generic statements which hold almost everywhere and so we ignore these technicalities.

Remark: At this point, nothing else in this chapter will be used later in the monograph.

5.7. The Pentagram Invariants

Here I will explain the pentagram invariants, which I first constructed in [**S3**]. By now there are many constructions. See e.g. [**Sol**] or [**GSTV**].

The pentagram map is well defined on $\mathcal{Z}_N$ and the equations given in equation (5.2) work in this more general setting. Indeed, these equations are even better suited to $\mathcal{Z}_N$ than to the smaller space $\mathcal{P}_N$ or ordinary N-gons. One can view $\mathcal{P}_N$ as a a codimension 8-subvariety of $\mathcal{Z}_N = \boldsymbol{R}^{2N}$.

If Ψ is a twisted N-gon, then $T(\Psi)$, the image of Ψ under the pentagram map, is another twisted N-gon with the same monodromy. Thus, the conjugacy class of the monodromy is a pentagram-invariant function on $\mathcal{Z}_N$.

The space $\mathcal{Z}_N$ admits an operation which is commonly called *scaling*. The scaling operation does not preserve $\mathcal{P}_N$ but nevertheless it is a key construction. Here is the scaling operation:

$$R_t(x_1, x_2, x_3, x_4, \dots) = (tx_1, t^{-1}x_2, tx_3, t^{-1}x_4, \dots). \tag{5.6}$$

R_t commutes with T^2. That is, $R_t \circ T^2 = T^2 \circ R_t$. The variable t is now sometimes called *the spectral parameter*.

Letting $x_1, \dots, x_{2N}$ be the flag invariants of P, define

$$E_N(P) = x_2x_4 \dots x_{2N}, \qquad O_N(P) = x_1x_3 \dots x_{2N-1}. \tag{5.7}$$

From equation (5.2) we see that $E_N(T(P)) = O_N(P)$ and $O_N(T(P)) = E_N(P)$. Hence E_N and O_N are both invariants of T^2.

Let M be the monodromy of a twisted polygon P. We lift M to an element of $GL_3(\boldsymbol{R})$ which we also denote by M. We define

$$\Omega_1 = \frac{\text{trace}^3(M)}{\det(M)}; \qquad \Omega_2 = \frac{\text{trace}^3(M^{-1})}{\det(M^{-1})}. \tag{5.8}$$

These quantities are independent of the lift of M and only depend on the conjugacy class of M. Hence they are invariants of T. The functions

$$\widetilde{\Omega}_1 = O_N^2 E_N \Omega_1; \qquad \widetilde{\Omega}_2 = O_N E_N^2 \Omega_2. \tag{5.9}$$

are invariants of T^2.

A polynomial in the flag coordinates has *weight* k if $R_t^*(P) = t^k P$. Here R_t^* denotes the obvious action of the scaling map R_t on polynomials. In [**S3**] we show that

$$\widetilde{\Omega}_1 = \left(1 + \sum_{k=1}^{(N-1)/2} O_k\right)^3; \qquad \widetilde{\Omega}_2 = \left(1 + \sum_{k=1}^{(N-1)/2} E_k\right)^3, \tag{5.10}$$

where O_k has weight k and E_k has weight $-k$. Since the functions $\widetilde{\Omega}_1$ and $\widetilde{\Omega}_2$ are invariants of T^2, and T^2 commutes with scaling, the polynomials $O_1, \ldots, O_n$, $E_1, \ldots, E_n$ are invariants of T^2. Here $n = (N-1)/2$. Combining these with O_N and E_N, we get $N+1$ invariants of T^2. More precisely, if we choose either sensible labeling scheme for the action of T, we get $O_k = E_k \circ T$ and $E_k = O_k \circ T$ for all k. See [**S3**] for a a combinatorial description of these polynomials.

5.8. Symplectic Manifolds and Torus Motion

A *symplectic manifold* is a smooth manifold M equipped with a closed, non-degenerate 2-form ω. What we mean is that $d\omega = 0$ and that $\omega \wedge \cdots \wedge \omega$ is a nondegenerate volume form. At each point $p \in M$, the form ω gives a canonical linear map between the tangent space T_pM and the cotangent space $(T_pM)^*$. The vector V maps to the linear functional X_V which has the property that

$$X_V(W) = \omega_p(V, W), \qquad \forall W \in T_p(M) \tag{5.11}$$

The nondegeneracy of M forces this map to be an isomorphism.

Given a smooth function $f : M \to \boldsymbol{R}$, we get a vector field, X_f, called *the Hamiltonian*, which has the property that $\omega(X_f, W) = df(W)$ for all vector fields W. In other words, we use the linear isomorphism to convert the 1-form df to the vector field X_f. Note that

$$df(X_f) = \omega(X_f, X_f) = 0. \tag{5.12}$$

Hence X_f is tangent to the level sets of f.

Let $C_\infty(M)$ denote the space of smooth functions on M. There is a natural bracket operation on $C_\infty(M)$, defined as follows:

$$\{f, g\} = \omega(X_f, X_g). \tag{5.13}$$

This bracket is a special case of what is called a *Poisson bracket*, which we will discuss below. One important property of the bracket is that

$$X_{\{f,g\}} = [X_f, X_g]. \tag{5.14}$$

The right hand side is the Lie bracket of vector fields, an operation that only depends on the smooth structure of M. In particular, if $\{f, g\} = 0$ then X_f and X_g generate commuting flows. In this situation, we say that f and g *Poisson commute*.

Suppose now that M is a compact symplectic manifold of dimension $2k$ and $f_1, \ldots, f_k$ are functions on M which pairwise Poisson commute. Suppose also that the differentials $df_1, .., df_k$ are almost everywhere linearly independent. At each $p \in M$ we have the *level set through* p, consisting of those points $q \in M$ such that $f_j(q) = f_j(p)$ for all j. Call L_p *clean* if $df_1, \ldots, df_k$ are linearly independent at every point of L_p. It follows from Sard's Theorem that almost every p is contained in a clean level set.

Let L be a clean level set. Since $df_1, \ldots, df_k$ are linearly independent at each point of L, the vector fields $X_1, \ldots, X_k$ are also linearly independent at each point of L. Moreover, these vector fields commute. In other words, L has a framing by commuting vector fields. Let's explore the topological consequences of this.

A *translation manifold* is a smooth k-dimensional manifold having coordinate charts into $\boldsymbol{R}^k$ so that the overlap functions are all restrictions of translations. Note that a translation manifold inherits the metric from $\boldsymbol{R}^k$ and, in particular, is locally isometric to $\boldsymbol{R}^k$. A compact translation manifold must be a torus, because it is universally covered by $\boldsymbol{R}^n$ and the covering group consists of translations. Our framed level set L has precisely this structure. We get the translation structure on M by integrating the vector fields. The fact that they commute gives us well-defined maps to $\boldsymbol{R}^k$. The overlap functions are all translations. Thus, each clean level set has a canonical metric in which it is isometric to a flat torus.

A *symplectomorphism* is a map $\phi : M \to M$ which preserves the symplectic form. That is, $\omega(V, W) = \omega(d\phi(V), d\phi(W))$ for all vector fields V and W. A *discrete completely integrable system* on a compact $2k$-dimensional symplectic manifold M is a symplectomorphism $\phi : M \to M$ which has k independent and pairwise Poisson-commuting invariant functions $f_1, \ldots, f_k$. In particular, $df_1, \ldots, df_k$ are almost everywhere linearly independent. In this situation, ϕ preserves all the level sets. Moreover, if L is a clean level, then L has a ϕ-invariant framing by commuting flows. This means that ϕ must be a translation relative to the translation structure on L. In other words, L has a flat metric relative to which ϕ is a translation. We say that ϕ exhibits *torus motion* on L. This works for almost all the level sets. Hence, almost every point of M exhibits torus motion under the action of ϕ.

The kind of integrability discussed above is known as *Arnold-Liouville integrability*. As we will see in the next section, this notion of integrability is defined in the wider context of Poisson manifolds.

5.9. Complete Integrability

In [**OST1**] we proved that the pentagram map is a discrete completely integrable system acting on $\mathcal{Z}_N$, in a sense which slightly broadens the notion of complete integrability discussed in the last section. I will sketch the proof here. For ease of exposition, I will take N odd. The even case is essentially the same, but the dimension count is a bit different. There are two extra invariants in the even case.

Let M be a smooth manifold and let $C_\infty(M)$ denote the space of smooth functions on M. A *Poisson bracket* is a map $\{\ ,\ \} : C_\infty(M) \times C_\infty(M)$ which satisfies the following properties.

- anti-symmetry: $\{f,g\} = -\{g,f\}$.
- linearity: $\{f_1 + f_2, g\} = \{f_1, g\} + \{f_2, g\}$.
- Liebniz rule: $\{f, gh\} = \{f,g\}h + \{f,h\}g$.
- Jacobi identity: $\{f,\{g,h\}\} + \{h,\{f,g\}\} + \{g,\{h,f\}\} = 0$.

A symplectic form on M gives rise to a Poisson bracket, as described in §5.8. The converse is not quite true. For instance, the 0-bracket does not come from a symplectic form. However, there is a close connection: The Poisson bracket defines a fibration of M by symplectic manifolds. (In the 0-bracket case, the fibers are individual points; this case is boring.)

There is a Poisson bracket on $\boldsymbol{R}^{2N}$ which has the following action on the coordinate functions

$$\{x_i, x_{i+2}\} = (-1)^i x_i x_{i+2}, \qquad \{x_i, x_j\} = 0 \text{ if } |i-j| \neq 2. \tag{5.15}$$

The bracket is extended to all rational functions using the Liebniz rule and linearity, and to all smooth functions using Taylor series. By direct calculation, we prove 3 algebraic facts about the Poisson bracket and the associated Hamiltonian derivative:

1. $\{\ ,\ \}$ is T-invariant. That is $\{f \circ T, g \circ T\} = \{f,g\} \circ T$ for all smooth functions. It suffices to check this on the coordinate functions.
2. $\{E_i, E_j\} = \{O_i, O_j\} = \{E_i, O_j\} = 0$ for all indices i, j.
3. $\{f, E_N\} = \{f, O_N\} = 0$ for all smooth functions f. That is, E_N and O_N are *Casimirs* for the bracket.

We also prove that the Poisson bracket, when restricted to the joint level sets of O_N and E_N, is non-degenerate. This, together with Sard's Theorem, implies that the generic level set of O_N and E_N is a symplectic manifold, equipped with a symplectic form that induces the above Poisson bracket. Whenever we have a compact level set of the remaining invariants, corresponding to a regular value of the map

$$p \to (O_1(p), \ldots, O_n(p), E_1(p), \ldots, E_n(p)),$$

we get the torus motion described in §5.8. In [**OST**] we isolate a nice class of twisted polygons, called *universally convex*, which undergo torus motion by virtue of lying on compact level sets.

So far we have been talking about the integrability of the pentagram map on the larger space $\mathcal{Z}_N$, whereas the reader might be specially interested in what happens on the smaller space $\mathcal{P}_N$. In [**OST2**] we proved that almost every point of $\mathcal{C}_N$ undergoes torus motion under some power of the pentagram map. We show that the Hamiltonian vector fields corresponding to the pentagram invariants on $\mathcal{Z}_N$ are everywhere tangent to $\mathcal{P}_N$ and that the span of these vector fields generically has the same dimension as the level sets of the invariants intersected with $\mathcal{P}_N$. These facts combine with Sard's theorem and the compactness of the level sets in $\mathcal{C}_N$ to guarantee the torus motion on $\mathcal{C}_N$ without quite getting all the classic features associated to integrability – e.g. an invariant Poisson bracket on $\mathcal{C}_N$.

Fedor Soloviev gave a proof in [**Sol**] that the pentagram map is integrable on $\mathcal{P}_N$, in the algebro-geometric sense, and this also implies the torus motion result for $\mathcal{C}_N$. The pentagram map is now known to be a special case of the spin networks considered in [**GSTV**], a special case of the cluster algebras considered in [**GP**], and a special case of the integrable systems defined in [**KS**], [**Bef1**], [**Bef2**], [**GK**], and [**F**].

CHAPTER 6

Some Related Dynamical Systems

In this chapter we discuss some dynamical systems related to the projective heat map.

6.1. Julia Sets of Rational Maps

Here I will give some very basic material on the notion of a Julia set for a single (complex) variable rational map. I include this section just so that the reader can see the classical definition of a Julia set and compare it to the definition of our set $\mathcal{J}$. This section barely scratches the surface of this vast topic. For much more information, see J. Milnor's book [**Mil**].

First we'll consider the case of polynomials. Let $n \geq 2$ and let

$$P(z) = a_n z^n + \cdots + a_1 z + a_0 \tag{6.1}$$

be a polynomial of degree n. The notation P^m denotes the m-fold composition of P with itself. Let U_P denote the set of points z such that

$$\lim_{m \to \infty} P^m(z) = \infty.$$

LEMMA 6.1. *U_P is an open set which contains a neighborhood of ∞.*

Proof: There is some constant C so that $|z| > C$ implies that $|P(z)| > 2|z|$. Hence, U_P contains a neighborhood of ∞. Also, if $z \in U_P$, then there is some n such that $|P^m(z)| > C$. But then $|P^m(w)| > C$ for all w sufficiently close to z. Hence U_P contains an open neighborhood about z. $\square$

Certainly $P(U_P) \subset U_P$ and $P^{-1}(U_P) \subset U_P$, and from these equations it is not hard to see that $P(U_P) = U_P$ and $P^{-1}(U_P) = U_P$. That is, U_P is *fully invariant*. The *Julia set* J_P has a simple definition: It is the boundary of U_P. Since U_P is open, J_P does not intersect U_P. Hence, every point in J_P has a bounded orbit. Moreover, J_P is bounded. Hence, J_P is compact. Since U_P is fully invariant, so is J_P. The Julia set is always a compact set without isolated points.

Remark: This definition of J_P lines up well with our set $\mathcal{J} = \mathcal{J}_5$, which we are calling the Julia set of the projective heat map H acting on $\mathcal{P}_5$. We will see that the set $\mathcal{U}$ of projective classes x such that $\{H^n(x)\}$ converges to the regular class is open. This follows from the easy fact that the regular class is an attracting fixed point of H. Thus, our set $\mathcal{J}$ is contained in the boundary of $\mathcal{U}$. We will see that $\mathcal{J}$ has measure zero, and in particular no interior. Hence $\mathcal{J}$ is exactly the boundary of $\mathcal{U}$. So, the definition is quite analogous to definition of the Julia set of a polynomial.

Julia sets in the quadratic case have been very well studied. In the quadratic case, one typically makes a change of variables so that $P(z) = z^2 + a$, and then considers the entire family as $a \in \boldsymbol{C}$ varies. The famous *Mandlebrot set* consists of those values $a \in \boldsymbol{C}$ for which J_P is connected.

Now we'll consider the case of rational maps. A single variable rational map is an expression of the form $f = P/Q$, where P and Q are polynomials. The simplest examples are the case when P and Q are linear maps. In this case f is a linear fractional transformation, or Mobius transformation. This case is rather boring from the point of view of complex dynamics.

Let $S^2 = \boldsymbol{C} \cup \infty$ denote the Riemann sphere. Let $U \subset S^2$ be some open set. A family $\mathcal{F}$ of holomorphic maps from U into S^2 is a *normal family* if every sequence in $\mathcal{F}$ has a subsequence which converges, uniformly on compact subsets to another holomorphic map on U. Here are some examples, all taking U to be the open unit disk.

- If $f(z) = z/2$ then the family $\{f^n\}$ forms a normal family. Any limiting map is either a contraction by a factor of 2^{-n} or the 0-map

- If $f(z) = z + 1$, then the family $\{f^n\}$ forms a normal family. Any limiting map is either a translation or the map which sends all points to ∞.

- If $f(z) = uz$ for some unit complex number z then the family $\{f^n\}$ is a normal family. Any limiting map is a rotation.

- If $f(z) = 2z$ then $\{f^n\}$ is not a normal family. The sequence itself has no convergent subsequence.

Given a rational map $f : S^2 \to S^2$, the *Fatou set* is defined to be the set of points $p \in S^2$ such that there exists an open neighborhood U of p such that the family of iterates of $f|_U$ forms a normal family. The *Julia set* J_f is the complement of the Fatou set.

Here is another characterization of the Julia set J_f. Let z be a periodic point for P. Say that $P^m(z) = z$. Setting $f = P^m$, we call z *repelling* if $|f'(z)| > 1$. From the discussion of normal families above, it is not hard to see that J_f must contain all the repelling periodic points. It turns out that the repelling periodic points are dense in J_f. So, one can characterize J_f as the closure of the set of repelling periodic points.

Remark: One can use normal families to define the Julia set of a rational map on $\boldsymbol{CP}^n$. For the case $n = 2$ see [**BDM**] for instance. We can define the projective heat map over $\boldsymbol{C}$, and perhaps the general notion of a Julia set for a rational map coincides with the closure of the set of projective classes of polygons which do not converge to the regular class upon iteration.

6.2. The One-Sided Shift

6.2.1. Basic Definition. The construction of the one-sided shift is based on some finite set F. For concreteness, we take $F = \{1, 2, 3, 4, 5, 6\}$. This is the example that comes up in connection with Theorem 1.5. The *shift space* is the space of all infinite sequences $\{a_i\}_{i=0}^{\infty}$ with $a_i \in F$. Call this space Σ_F. This space is naturally a metric space. Define the distance from $\{a_i\}$ to $\{b_i\}$ to be 6^{-K} where

K is the smallest integer such that $a_K \neq b_K$. The space Σ_F has diameter 1, and is homeomorphic to a Cantor set. The *one-sided shift* is the map $\phi : \Sigma_F \to \Sigma_F$ defined by: $\phi(\{a_i\}) = \{b_i\}$ where $b_i = a_{i+1}$ for all i. In other words ϕ just chops the zeroth term off the sequence. By definition, ϕ expands distances by a factor of 6 and is 6-to-1.

A *periodic point* in Σ_F is a point p such that $\phi^n(p) = p$ for some n. The smallest n for which this holds is called the *period* of the point. This definition makes sense in any dynamical system.

LEMMA 6.2. *Σ_F has periodic points of all orders, and the periodic points are dense in Σ_F.*

Proof: Any sequence in Σ_F which repeats after n steps is a periodic point of period n. Hence Σ_F has periodic points of all orders. Any point in Σ_F can be approximated arbitrarily well by a periodic point. We just take the first n terms of the point and then make it repeat endlessly. Hence the periodic points are dense. □

Σ_F naturally has a measure: The measure of the set of all sequences starting $i_0, \ldots, i_k$ has measure 6^{-k}. These sets are called *cylinder sets*. Equipped with this measure, ϕ is a measure-preserving map in the sense that $\phi^{-1}(S)$ and S have the same measure for any measurable subset of Σ_F. It suffices to check this on the cylinder sets. The inverse image of a cylinder set of size 6^{-k} is a disjoint union of 6 cylinder sets of size 6^{-k-1}.

A subset $S \subset \Sigma_F$ has measure 0 if, for every $\epsilon > 0$, we have

$$S \subset \bigcup C_i, \qquad \sum \mu(C_i) < \epsilon.$$

Here $\{C_i\}$ is a countable collection of cylinder sets.

LEMMA 6.3. *Almost every point of Σ_F has a dense orbit.*

Proof: A point in Σ_F has a dense orbit provided that it contains every finite sequence. We will show that the set of points which do not have a particular finite sequence has measure zero. Since the countable union of sets of measure zero has measure zero, the set of points which avoid some subsequence has measure zero. Hence, almost all points contain all finite sequences.

Let S be a finite sequence. Suppose that S has length n. Let $\Sigma_F(S)$ denote the set of points which do not have S. Suppose that S has length n. Let $\Sigma'_F(S)$ denote the set of points with the following property: There is no k for which $a_{kn}, \ldots, a_{(k+1)n-1}$ is S. In other words, elements of $\Sigma'_F(S)$ might contain S but just not in positions starting at $0, n, 2n, 3n, \ldots$.

Obviously $\Sigma_F(S) \subset \Sigma'_F(S)$. So, it suffices to prove that $\Sigma'_F(S)$ has measure 0. Note that $\Sigma'_F(S)$ is contained in $6^n - 1$ cylinder sets of size 6^{-n}. Each of these cylinder sets intersects $\Sigma'_F(S)$ in exactly $6^n - 1$ cylinder sets of size 6^{-2n}, and so on. Hence, for any k, we see that $\Sigma'_F(S)$ is contained in a union of $(6^n - 1)^k$ cylinder sets of size 6^{-kn}. The total measure of these cylinder sets is less than any desired ϵ provided that k is taken large enough. □

6.2.2. The Shift in Action. Now we show the 1-sided shift arises in a 2-dimensional context. Suppose K is a closed topological disk and $K_1, \ldots, K_6 \subset D$ are 6 pairwise disjoint topological disks. Suppose we have a rational map h on $\boldsymbol{R}^2$ such that

- For $k = 1, \ldots, 6$, we have $h(K_k) = D$ and the restriction of h to some open neighborhood of K_k is a diffeomorphism.
- For every point of $D - K_k$ where h is defined, we have $h(p) \notin D$.

We define $\heartsuit K \subset K_1 \cup \cdots \cup K_6$ to be those p such that $h^n(p) \in D$ for all n. To analyze this set, we let $K(n)$ denote those points p such that $h^n(p) \in D$. Evidently

$$\heartsuit K = \bigcap K(n).$$

Each disk K_j contains 6 smaller topological disks $K_{ij} = K_i \cap h^{-1}(K_j)$. Each of the 36 disks K_{ij} contains 6 smaller disks $k_{ijk} = K_i \cap h^{-1}(K_{jk})$. And so on. From this description we see that $K(n)$ consists of 6^n pairwise disjoint disks, each of which contains 6 disks of $K(n+1)$.

To get more information about $\heartsuit K$ we need some control over the sizes of these disks. The assumption we make is that there exists some $n_0 \geq 1$ and some $\eta > 1$ such that the restriction of h to each disk of $K(n_0)$ expands distances by at least η. From this assumption we see that any nested family of disks in the intersection $\bigcap K(n)$ shrinks to a point.

Thus, we recognize $\heartsuit K$ as a Cantor set. To be more precise, we have a well-defined map $\Psi : \Sigma_F \to \heartsuit K$. Each point in the shift space Σ_F corresponds to a nested sequence of disks in this construction and Ψ maps this point to the intersection of the corresponding disks. Ψ is surjective and, thanks to the expanding property, continuous. Any continuous surjective map from a compact space to a subset of the plane is a homeomorphism. Hence Ψ is a homeomorphism. Moreover, by construction Ψ is a *conjugacy*: $\Psi\phi\Psi^{-1} = h$. Thus, we recognize the action of h on the set $\heartsuit K$ as (conjugate to) the one-sided shift on 6 symbols.

6.2.3. A Contrived Generalization. When it comes time to consider an application of the analysis above in §12.5 we will have a slightly more contrived situation. We explain it here. We assume that we have a piecewise smooth and everywhere continuous function ρ on the tangent bundle of D that is comparable to the Euclidean metric in the sense that there is some uniform constant $C > 0$ such that $C^{-1}\|V\| \leq \rho(V) \leq C\|V\|$ for all vectors V. We call ρ a *tangent bundle function*.

We say that h is η-expanding on a set S, with respect to ρ, if

$$\rho_{h(p)}(dh(V)) \geq \eta\, \rho_p(V), \tag{6.2}$$

for all $p \in S$ and all tangent vectors V based at p. We get the same shrinking disks conclusion if we assume that there is some n_0 and some $\eta > 1$ so that h is η-expanding on each disk of $K(n_0)$ with respect to ρ.

Here's the proof. Choose some arbitrary disk κ_n of $K(n_0 + n)$ and consider $\kappa_0 = h^n(\kappa_n)$. The disks $h(\kappa_n), \ldots, h^n(\kappa_n) = \kappa_0$ all lie in $K(n_0)$. The function ρ allows us to define the lengths of paths in D, and by the expansion property, the length of any path in κ_n is at most η^{-n} as long as its image in κ_0. This combines with the comparison to the Euclidean metric to show that the Euclidean diameter of κ_n is at most $C\eta^{-n}$ for some constant C that does not depend on any choices.

6.3. The Two-Sided Shift

The two-sided shift is based on some finite set F. We define the space Θ_F to be the set of bi-infinite sequences $\{a_i\}_{i=-\infty}^{\infty}$. The metric on Θ_F is defined in a similar way to the one on Σ_F. Here $d(\{a_i\},\{b_i\}) = |F|^{-K}$, where K is the smallest integer such that $a_k = b_k$ for all $|k| < K$. The space Θ_F is compact.

The two-sided shift is the map $f(\{a_k\}) = \{b_k\}$ where $b_k = a_{k+1}$. The map f is a homeomorphism from Θ_F to itself. As with the one-sided shift, Θ_F has a natural measure with respect to which f is measure preserving. f has periodic points of all orders, and the periodic points are dense, and almost every point has a dense orbit. The proofs are essentially the same as for the one-sided shift. When we look at the Smale Horseshoe, we will see how the two-sided shift arises naturally in the context of a planar diffeomorphism.

6.4. The Smale Horseshoe

Let $X = [0,1] \times [-1/2, 3/2]$. The Smale horseshoe is a smooth map $f : X \to X$ which has the topological features shown in Figure 6.1.

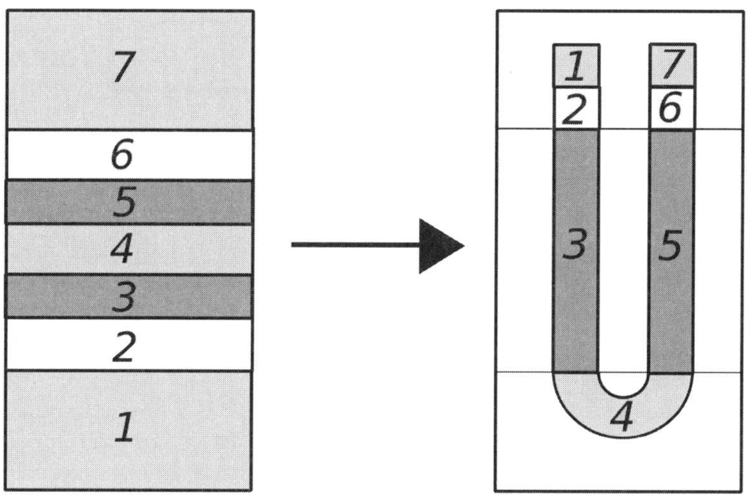

FIGURE 6.1. The Smale horseshoe

The map also has the following additional properties.

- $f : X \to f(X)$ is a diffeomorphism.
- f is a contraction on X_1 and on X_7. In particular, f has an attracting fixed point $p_\infty \in X_7$.
- f is an affine map on X_3 and on X_5. Here the derivative df is a diagonal matrix of the form

$$\begin{bmatrix} \lambda^{-1} & 0 \\ 0 & \lambda \end{bmatrix} \qquad \lambda > 1.$$

Note that $f^3(p) \in X_7$ when $p \in X - X_3 - X_5$, and then for such points $f^n(p) \to p_\infty$. So, the only points not attracted to p_∞ under iteration are the points p such that $f^n(p) \in X_3 \cup X_5$ for all n. Let A denote the set of these points.

Let $X' = X_3 \cup X_5$. The set of points $p \in X'$ such that $f(p) \in X'$ is the union of 4 rectangles. These rectangles have the form $X_{ij} = X_i \cap f^{-1}(X_j)$ for $i, j \in \{3, 5\}$. These 4 rectangles are shown in Figure 6.2.

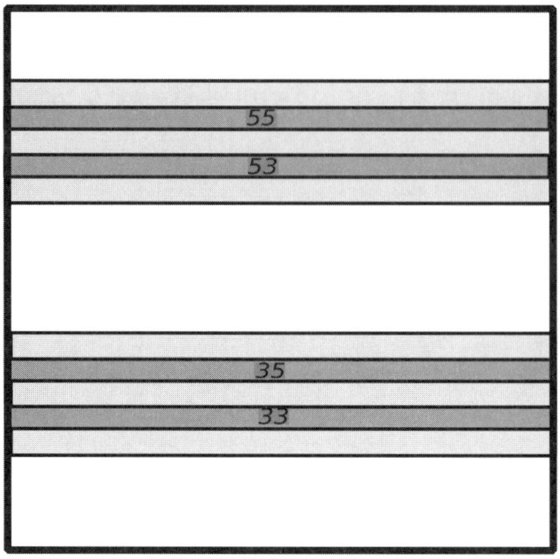

FIGURE 6.2. X_{33} and X_{35} and X_{53} and X_{55}

The pattern continues, the set $p \in X$ such that $f^k(p) \in X'$ for $k = 0, \ldots, n-1$ consists of 2^n rectangles, indexed by all sequences of length n in $F = \{3, 5\}$. The intersection A of all these sets is $\Sigma_F \times [0, 1]$, the product of a Cantor set and an interval. The action of f on X_s maps horizontal line segments to horizontal line segments, and the action on these line segments is the 1 sided shift on Σ_F.

The inverse map f^{-1} is defined on $Y_3 = f(X_3)$ and $Y_5 = f(X_5)$. (These are the shaded rectangles on the right side of Figure 6.1.) The same analysis shows that the set B of points $p \in Y_3 \cup Y_5$ such that $f^{-n}(p) \in Y_3 \cup Y_5$ for all n is the product $[0, 1] \times \Sigma_F$, and the action of f^{-1} on the vertical line segments of B is the 1-sided shift. B is just A turned sideways. The set B is the attracting set for the restriction of f to A. For every $p \in A$, the forward orbit $\{f^n(p)\}$ accumulates on $A \cap B$. Likewise, $A \cap B$ is the attracting set for the restriction of f^{-1} to B. Figure 6.3 hints at the nature of the set $A \cap B$.

Each point $p \in A \cap B$ has two infinite sequences attached to it. The sequence s_A describes the position of p in A and the sequence s_B describes the sequence in B. form the bi-infinite sequence $s_B.s_A$, where s_B is written right to left and s_A is written left to right. The decimal point indicates that term 0 of this bi-infinite sequence is the first term in s_A. The map f moves the decimal point one unit to the right and the map f^{-1} moves the decimal point one unit to the left. Thus, the restriction of f to $A \cap B$ is conjugate to the 2-sided shift.

6.5. Quasi Horseshoe Maps

Here I describe the kind of map that appears in Theorem 1.6, something I call a *quasi-horseshoe*. The notion of a quasi-horseshoe is a relaxation of the notion of

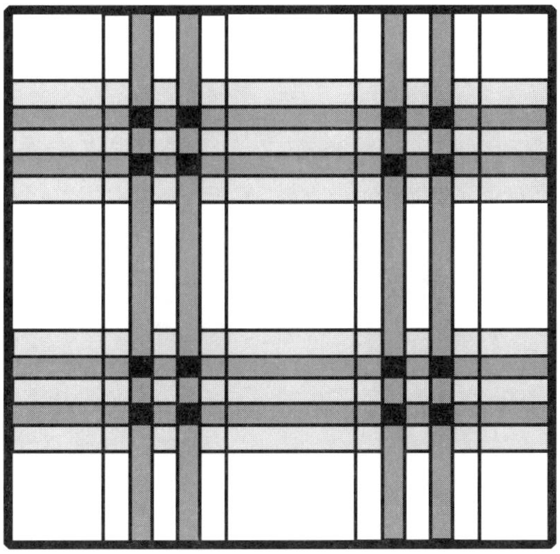

FIGURE 6.3. A hint of the set $A \cap B$

a Smale horseshoe. The main point of making the relaxation is that it is easier to establish the existence of a quasi-horseshoe than it is to establish the existence of a Smale horseshoe. I do not know if my definition of a quasi-horseshoe arises in the literature.

6.5.1. Adapted Quadrilaterals. Let $\vee \subset \boldsymbol{R}^2$ denote the standard light cone. Here $\vee$ is the set of vectors in $\boldsymbol{R}^2$ whose slope exceeds 1 in absolute value. We say that a curve is *timelike* if all of of its chords lie in $\vee$. (A chord of the curve is a vector pointing from one point on the curve to another.) We say that a curve is *spacelike* if reflection in the diagonal maps it to a timelike curve. These are the usual definitions.

We say that an *adapted quadrilateral* (or *quad* for short) is a piecewise smooth embedded loop with 4 distinguished vertices we call *corners*, such that one pair of opposite sides is timelike and the other pair is spacelike. Here a *side* is an arc of the quad connecting two consecutive corners. We say that two quads are *interlaced* if the following holds.

- Each spacelike edge of one quad crosses each timelike edge of the other.
- The spacelike edges of one quad are disjoint from the spacelike edges of the other.
- The timelike edges of one quad are disjoint from the timelike edges of the other.

Figure 6.4 shows a quad, two interlaced quads, and one quad interlacing four others.

6.5.2. Interlacing and Quasi-Hyperbolicity. Let P_0 be a quad and let $P_1, \ldots, P_k$ be a finite list of quads. We say that a continuous map $F : P_0 \to \boldsymbol{R}^2$ *interlaces* $P_1, \ldots, P_k$ if

(1) $F(P_0) \subset P'$, where P' is a quad which interlaces $P_1, \ldots, P_k$.

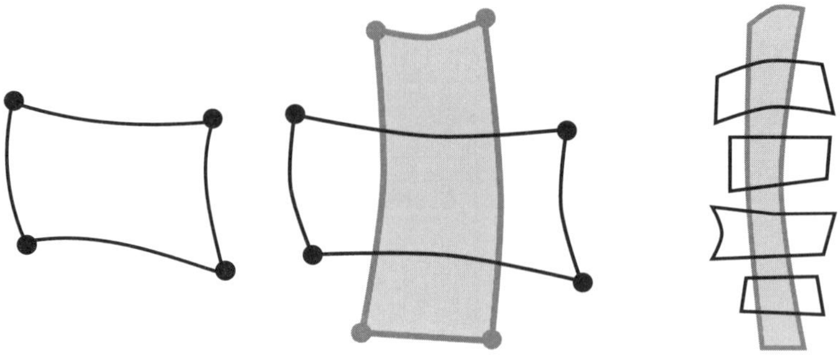

FIGURE 6.4. A quad and some interlaced quads

(2) F maps one of the spacelike edges of P_0 above $P_1, \ldots, P_k$ and one below. In each case, we mean that some horizontal line separates the relevant sets.

Figure 6.5 shows what we mean.

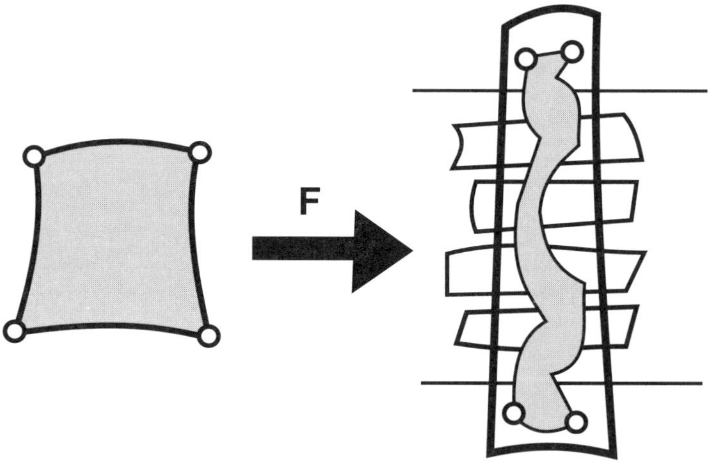

FIGURE 6.5. $F(P_0)$ interlaces P_1, P_2, P_3, P_4

Now we add some more structure. Given $\lambda > 1$, we call F *λ-quasi-hyperbolic* on P_0 if

- F is a local diffeomorphism in a neighborhood of P_0.
- $dF(\vee) \subset \vee$ strictly at all points in a neighborhood of P_0.
- $\|dF(V)\| \geq \lambda \|V\|$ for all $V \in \vee$.

We call λ the *stretch factor*. When the choice of λ is not important, we simply say that F is *quasi-hyperbolic*.

Remark: Being quasi-hyperbolic is weaker than what people usually mean by *hyperbolic*. In the hyperbolic case, one would want dF to have one eigenvalue less than 1 and one eigenvalue greater than 1.

6.5.3. A Preliminary Definition. I hope the reader will forgive the upcoming terminology. I first want to describe a map which has many of the features of a

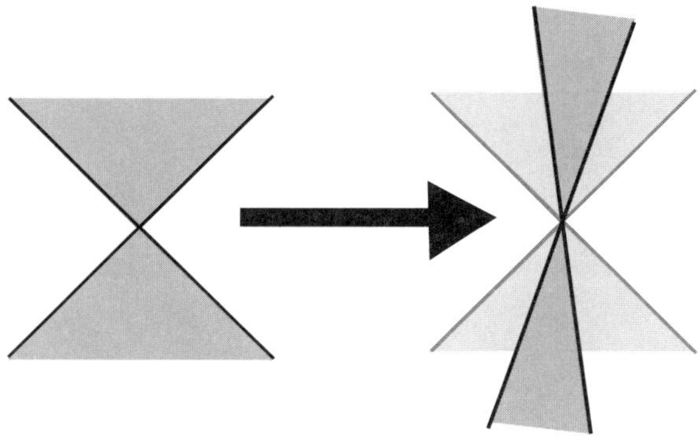

FIGURE 6.6. The cone $\vee$ and its image $dF(\vee) \subset \vee$

quasi-horseshoe but lacks one complicating definition. Most of the analysis we care about does not depend on the extra definition.

Let $\Omega = P_1 \cup \cdots \cup P_k$. Here $P_1, \ldots, P_k$ are pairwise disjoint quads. We say that a *quasi quasi-horseshoe* is a map $F : \Omega \to \boldsymbol{R}^2$ with the following properties.

- $F(P_j)$ interlaces $P_1, \ldots, P_k$ for each j.
- The restriction of F to each P_j is λ quasi-hyperbolic for some $\lambda > k$ that does not depend on the index j.

Let $A \subset \Omega$ be the set of points x such that $F^n(x) \in \Omega^o$ for all $n = 0, 1, 2, \ldots$. Note that $A \subset \Omega^o$. Here Ω^o is the interior of Ω.

LEMMA 6.4. *A has (2-dimensional Lebesgue) measure* 0.

Proof: Let L_1 denote length. Consider the family $\mathcal{T}$ of all smooth timelike curves which intersect A. Let M be the supremum of $L_1(\gamma \cap A)$, where $\gamma \in \mathcal{T}$. Certainly $M < 2\,\mathrm{diam}(\Omega)$. If A has positive measure then, by Fubini's Theorem, we can find a vertical line which intersects A in a set of positive length. Hence $M > 0$.

Recall that $k/\lambda < 1$. Here λ is the stretch factor. Choose some $\gamma_1 \in \mathcal{T}$ such that $L_1(\gamma_1 \cap A) > (k/\lambda)M$. But then there is some index j such that

$$L_1(\gamma_1 \cap A \cap P_j) > M/\lambda.$$

Given the the interlacing property of F, we see $\gamma_2 = F(\gamma_1 \cap P_j) \in \mathcal{T}$. Given the λ-stretching property, and the forward-invariance of A, we see that $L_1(\gamma_2 \cap A) > M$. This is a contradiction. $\square$

Recall that a Cantor band is a space homeomorphic to the product of a Cantor set and an open interval. Note that a Cantor band is naturally a union of maximal embedded open arcs. We call these arcs the *strands* of the Cantor band.

LEMMA 6.5. *A is a Cantor band. All the strands of A are spacelike.*

Proof: Let A' be the set of points x such that $F^n(x) \in \Omega$ for all $n = 1, 2, 3, \ldots$. The difference between A' and A is that $A = A' \cap \Omega^o$. We will show that A' is homeomorphic to $[0,1] \times Y$, where Y is a Cantor set. We get A by removing the endpoints of all the arcs of A'.

Let $\Omega(0) = \Omega$ and let $\Omega(n+1) = F^{-1}(\Omega(n))$. Note that

$$A' = \bigcap_{n=0}^{\infty} \Omega(n). \tag{6.3}$$

The interlacing property combines with induction to show that $\Omega(n)$ consists of k^n quads, and each quad of X_{n-1} contains k quads of $\Omega(n)$. These quads are pairwise disjoint and have their timelike sides in the timelike sides of X_0. Figure 6.7 shows X_0 and X_1 when $k = 2$.

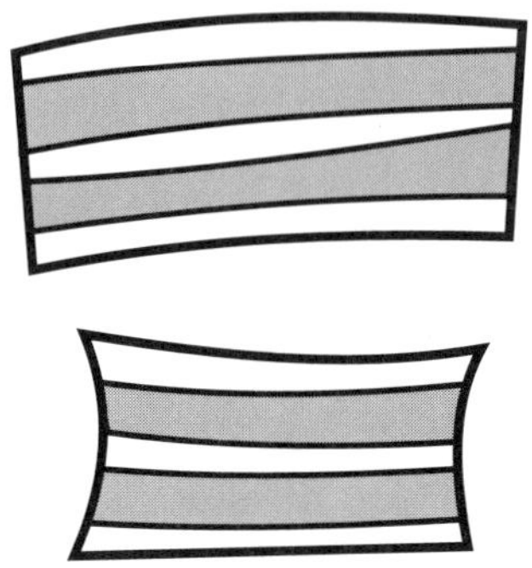

FIGURE 6.7. X_0 and X_1 when $k = 2$

Let Y_n denote a collection of k^n pairwise disjoint segments so that each segment of Y_n contains k segments of Y_{n-1}. Choose these segments so that $Y = \bigcap Y_n$ is a Cantor set – i.e. every infinite nested intersection is a point. There is a homeomorphism h_n between $\Omega(n)$ and $[0,1] \times Y_n$ which has constant speed along the spacelike edges of the quads in $\Omega(n)$.

The same argument as in the measure 0 case shows that each vertical line intersects each quad of $\Omega(n)$ in a segment of length at most $O((k/\lambda)^n)$. Hence, any infinite nested intersection of these quads is a single spacelike arc. From this property, our sequence of homeomorphisms induces a homeomorphism between A' and $[0,1] \times Y$.

Finally, there is a dense set of strands in A corresponding to the tops and bottoms of the nested quads. These strands are all spacelike. Hence, their limit strands are spacelike in the weak sense that their chords are never timelike. If such a limit strand γ has a chord of slope ± 1 then the image $F(\gamma)$ has a timelike chord by the quasi-hyperbolicity. Hence γ only has spacelike chords. Hence all strands in A are spacelike. □

We can get one more result without too much trouble. The basic idea behind the next result is that the quasi-hyperbolicity exaggerates a hypothetical point of non-differentiability (or non-continuous differentiability) to the point where it contradicts the spacelike nature of the arcs.

LEMMA 6.6. *The strands of A are continuously differentiable.*

Proof: As in the previous result, we work with the larger set A'. We will show that the strands of A' are differentiable. At the endpoints of the strands of A' we mean to use the one-sided definition of the derivative. We work with this larger space because it is compact. Exactly the same argument shows that the strands are in fact continuously differentiable.

Let α be the line through the origin of slope -1 and let δ be the line through the origin of slope 1. These are the two boundaries of the lightcone $\vee$. Suppose some arc μ of A' is not differentiable at some point $x \in \mu$. Then there are sequences of points $\{y_n\}$ and $\{x_n\}$, all in μ, so that $y_n \to x$ and $z_n \to x$ and the angles between the lines $\beta_n = (xy_n)$ and $\gamma_n = (xz_n)$ do not converge to 0. We label so that the slopes of $\alpha, \beta_n, \gamma_n, \delta$ come in order. Consider the cross ratios of the slopes:

$$[x] = \limsup_{n\to\infty} [\alpha, \beta_n, \gamma_n, \delta]. \tag{6.4}$$

By compactness, we can choose x so that $[x]$ is as large as possible.

Suppose we apply F to this whole picture. Let $x' = F(x)$, etc. Let β'_n be the line $(x'y'_n)$ and let γ'_n be the line $(x'z'_n)$. Recall that $dF(\vee)$ is contained strictly inside $\vee$. If $[x] = \infty$ then either $\beta_n \to \alpha$ or $\gamma_n \to \delta$. In either case, one of β'_n or γ'_n lies strictly in $\vee$ for n large. This contradicts the spacelike nature of the curves.

Now we know that $[x]$ is finite and moreover β_n does not converge to α and γ_n does not converge to δ. Hence, for n sufficiently large we have

$$[\alpha, \beta_n, \gamma_n, \delta] = [\alpha', \beta'_n, \gamma'_n, \delta'] \leq (1-\epsilon)[\alpha, \beta'_n, \gamma'_n, \delta], \tag{6.5}$$

for some $\epsilon > 0$ independent of n. This gives $[x] \leq (1-\epsilon)[x']$, a contradiction. □

6.5.4. Main Definition. Here we motivate our main definition by pointing out one shortcoming of the definition above. Let F be a quasi quasi-horseshoe as above. Let B denote the set of accumulation points of F-orbits in A. Let α be a strand of A. We might have a situation where F maps each P_j onto a thin neighborhood of a single timelike curve and contracts the strands of A in this neighborhood. In this case, B would intersect each strand of A in a single point. This seems rather unlike what happens with the horseshoe.

We say that F is a *quasi-horseshoe* if we can partition each P_j into a left and a right half

$$P_j = P_j^1 \cup P_j^2, \tag{6.6}$$

so that the following things are true.

(1) $F(P_j)$ interlaces $P_1^1, \ldots, P_k^1$ or $P_1^2, \ldots, P_k^2$ for each j.

(2) Each of the options in Item 1 occurs for some index.

What we are saying is that sometimes F maps the quads over the left half and sometimes F maps the quads over the right half.

Remark: Unlike the Smale horseshoe, a quasi-horseshoe need not be an injective map.

Given a subset $S \subset A$, let S^* denote the set of accumulation points of S. Inductively define $S^{(n)} = (S^{(n-1)})^*$. Say that S has *infinite Cantor-Bendixson rank* if $S^{(n)}$ is nonempty for all n.

LEMMA 6.7. *If F is a quasi-horseshoe, then B intersects each strand of A in a set of infinite Cantor-Bendixson rank.*

Proof: Since F is a local diffeomorphism and F maps each strand of A into a strand of A, the restriction of F to each strand is injective.

The action of F on the strands of A is conjugate to the one-sided shift on k symbols. In particular, this strand-action has dense orbits. Hence B intersects a dense set of strands of A. Recall that A' is the compact set obtained by adjoining the endpoints of each strand of A. Since B is closed and A' is compact, we see that B intersects every strand of A'. Since $F(A') \subset A$ we see, finally, that B intersects every strand of A.

But now we can apply F and use the definition above to say that B intersects each strand of A in two points, one on the left and one on the right. Now we can iterate. Since the restriction of F to each strand is injective, we can say that B intersects A in at least 2^n points for every n. Hence B intersects each strand of A in an infinite set of points.

Let α be a strand of A. Since $F(A') \subset A$ and each strand of A' is compact, we can see that $B \cap \alpha$ has at least one accumulation point in α. Hence B^* intersects every strand of A. But now we can repeat the same argument as above to show that B^{**} intersects every strand of A. And so on. □

Remark: One can cook up examples where B intersects the strands of A in sets which are not Cantor sets.

6.6. The 2-adic Solenoid

A 2-*adic* integer is an infinite sequence of the form $a_1, a_2, a_3, \ldots$ where $a_j \in \boldsymbol{Z}/2^j$ and

$$a_{j+1} \equiv a_j \mod 2^j \qquad \forall j. \tag{6.7}$$

The set $\boldsymbol{Z}_2$ of such sequences forms a ring. One does coordinatewise addition and multiplication in the corresponding finite rings and the compatibility given by equation (6.7) makes this well defined. We only care about the addition and not the multiplication in what we say below, though people interested in p-adic dynamics generally care about both.

The set $\boldsymbol{Z}_2$ also has a natural metric, the 2-*adic metric*. The distance between two points $\{a_k\}$ and $\{b_k\}$ is 2^{-N} where N is the smallest integer where the sequences disagree. This makes $\boldsymbol{Z}_2$ into a metric ring. There is a natural inclusion of $\boldsymbol{N}$, the natural numbers, into $\boldsymbol{Z}_2$. The point n maps to the sequence $\{a_j\}$ where a_j is the reduction of n mod 2^j. We simply think of $\boldsymbol{N}$ as a subset of $\boldsymbol{Z}_2$. The set $\boldsymbol{N}$ consists of those sequences which are eventually constant. Clearly $\boldsymbol{N}$ is dense in $\boldsymbol{Z}_2$. Hence $\boldsymbol{Z}_2$ is the completion of $\boldsymbol{N}$ with respect to 2-adic metric. With the 2-adic metric, the space $\boldsymbol{Z}_2$ is homeomorphic to a Cantor set.

The 2-*adic odometer* is the map $x \to x + 1$ acting on $\boldsymbol{Z}_2$. The orbit of 0 under this map is $\boldsymbol{N}$. Hence 0 has a dense orbit. The orbit of any other point under the 2-adic odometer is just a translate of $\boldsymbol{N}$. Hence, every point has a dense orbit.

The 2-adic solenoid is the mapping cylinder for the 2-adic odometer. That is, we take the space $[0, 1] \times \boldsymbol{Z}_2$ and we identify the points $(0, y)$ with $(1, y + 1)$. More canonically we take $\boldsymbol{R} \times \boldsymbol{Z}_2$ and we take quotient out by the $\boldsymbol{Z}$ action generated by the map $(x, y) \to (x + 1, y + 1)$.

The 2-adic solenoid is partitioned into infinite curves. Each infinite curve is dense in the solenoid. The M-fold cyclic cover of the 2-adic solenoid is the quotient

$\boldsymbol{R} \times \boldsymbol{Z}_2$ by the subgroup $M\boldsymbol{Z}$. Here $M\boldsymbol{Z}$ is generated by the map

$$(x, y) \to (x + M, y + M).$$

Theorem 1.10 mentions the 5-fold cover.

Finally, we mention that we can think of the 2-adic solenoid as a fibration over the circle. The fibration map is just the projection onto the first coordinate. The fibers (a.k.a. cross-sections) are copies of $\boldsymbol{Z}_2$ and hence homeomorphic to Cantor sets. Thus, one can picture the 2-adic solenoid as a kind of twisted version of the product of a circle and a Cantor set.

6.7. The BJK Continuum

Here is another point of view on the 2-adic integers. One can alternatively represent a 2-adic integer as a formal infinite series

$$b_0 + 2b_1 + 4b_2 + \ldots \tag{6.8}$$

Here the b_i are either 0 or 1. We get a 2-adic sequence by taking the partial sums $a_1 = b_0$ and $a_2 = b_0 + 2b_1$, and so on. This gives us an identification of $\boldsymbol{Z}_2$ with the set of infinite binary sequences. With this identification, addition is done the way one learns it in elementary school, except that one works in base 2 and carries to the right. We call this identification the *series identification*.

Remark: Here is how to see the homeomorphism between $\boldsymbol{Z}_2$ and the middle third Cantor set. We take the series in equation (6.8) and map it to the base 3 expansion

$$.(2b_0), (2b_1), (2b_2), \ldots$$

This expansion contains only 0s and 2s and hence defines a point in the middle third Cantor set.

We identify $\boldsymbol{Z}_2$ with the set of infinite binary strings, using the series identification. We consider the following two involutions on $\boldsymbol{Z}_2$. The involution I_0 reverses all the digits after the first 1 is encountered. For instance,

$$I_0(\mathbf{001}0101110\ldots) = \mathbf{001}1010001\ldots$$

The involution I_1 reverses all the digits.

We can use these involutions to construct the *Brouwer-Janiszewski-Knaster continuum*. This space is obtained as follows. We start with the $[0, 1] \times \boldsymbol{Z}_2$ and we identify the points (j, y) to $(j, I_j(y))$ for $j = 0, 1$. Figure 6.8 shows the first few steps in a concrete construction of the BJK continuum.

FIGURE 6.8. Constructing the BJK Continuum

Since $\boldsymbol{Z}_2$ is a ring, we can interpret the involutions I_0 and I_1 algebraically. We do this by working formally with the series representation of points in $\boldsymbol{Z}_2$.

LEMMA 6.8. *$I_1(x) = -1 - x$.*

Proof: Using the rule for addition mentioned above, we have

$$(1, 0, 0, 0, \dots) + (1, 1, 1, 1, \dots) = 0.$$

Here is what is going on: When we perform the addition, we get $1 + 1 = 0$ in the leftmost position and then we carry the 1 to the right. This gives 0 in the second position and again we carry the 1 to the right. And so on. The remainder of 1 sweeps rightward, leaving all 0s in its wake. What this means is that $(1, 1, 1, 1, \dots)$ is the series representation for -1. Now, given $x = (b_0, b_1, b_2, \dots) \in \boldsymbol{Z}_2$, we have

$$I(x) = (1 - b_0, 1 - b_1, 1 - b_2, \dots).$$

Adding coordinatewise, we get $x + I(x) = (1, 1, 1, \dots) = -1$. □

LEMMA 6.9. *$I_0(x) = -x$.*

Proof: Taking the series representation for x as in equation (6.8), we have

$$x + I_0(x) = 1(0) + 2(0) + \cdots + 2^k(0) + 2^{k+1}(1 + 1) + 2^{k+2}(1) + 2^{k+3}(1) + \cdots = 0.$$

Here $k + 1$ is the index of the first 1 in the series representation of x. What is going on here when we add is that we first produce k consecutive 0s and then we have $1 + 1$, which produces another 0 with a remainder of 1. Following this, the remainder sweeps rightward, leaving an infinite string of 0s in its wake. □

We use our algebraic knowledge of the maps I_0 and I_1 to define the BJK continuum in a different way. We let I_0 and I_1 act on $\boldsymbol{R} \times \boldsymbol{Z}_2$ by the diagonal action $I_0(x, y) = (-x, -y)$ and $I_1(x, y) = (-1 - x, -1 - y)$. Then $\langle I_1, I_2 \rangle$ is the infinite dihedral group and this group acts on $\boldsymbol{R} \times \boldsymbol{Z}^2$. The quotient space is the BJK continuum.

Observe that

$$(I_0 I_1)^M(x, y) = (x + M, y + M). \tag{6.9}$$

In this way we recognize the 2-adic solenoid as the double cover of the BJK continuum. Moreover, the M-fold cover of the 2-adic solenoid is an order $2M$ dihedral cover of the BJK continuum. This is how we will recognize the space from Theorem 1.10 in the case $M = 5$.

Part 2

CHAPTER 7

The Projective Heat Map

In this chapter we will derive formulas for the projective heat map in general, and then specialize to the case $N = 5$. The last section in this chapter contains some speculation about what the projective heat map does for $N \geq 6$, but really there are no results here. Following this chapter, the rest of the monograph is about the case $N = 5$.

7.1. The Reconstruction Formula

In general, if we want to derive the equation for some projectively natural iteration Ψ in terms of these flag invariants, from §3.6, we take the following 3 steps.

(1) Start with a list $(x_0, x_1, x_2, \ldots)$ of flag invariants and construct the polygon P which has these flag invariants.
(2) Compute the polygon $\Psi(P)$.
(3) Compute the flag invariants $(y_0, y_1, y_2, \ldots)$ of $\Psi(P)$.

The desired rational map is then $(x_0, x_1, x_2, \ldots) \to (y_0, y_1, y_2, \ldots)$.

In any given case, the description of the map Ψ would presumably give a recipe for doing Step 2. This is certainly true for the projective heat map. In §3.6 we give a recipe for Step 3. What really needs to be explained is Step 1.

Step 1 comes from work in [**S3**] where I give a general formula for P in terms of the list $(x_0, x_1, x_2, \ldots)$. Note that this list of flag invariants does not uniquely specify P. Any projectively equivalent polygon has the same list of flag invariants. My formula makes some specific choice of P by normalizing the first few vertices and edges in a certain way. I call this formula the *reconstruction formula*. I originally used the reconstruction formula to compute things about the pentagram map.

The derivation of the reconstruction formula in [**S3**] is rather long and complicated. However, for the purposes of deriving the formula for the projective heat map acting on $\mathcal{P}_N$, we only need to know the first 8 points of our polygon. The reader who does not want to think about the general reconstruction formula can simply plug the formulas given below for the 8 points

$$P_{-7}, P_{-3}, P_1, P_5, P_9, P_{13}, P_{17}, P_{21}$$

into a symbolic manipulator and see that they are indeed the correct first 8 points for P. We present the general reconstruction formula so that the interested reader can see where our formulas for the 8 above-mentioned points actually come from.

We make one more apology before presenting the reconstruction formula. The points are indexed in a funny way, but this way of indexing makes the general formula as clean as possible. Now for the formula.

Our polygon will have vertices P_{9+2k} for $k = -8, -6, -4, -2, \ldots$. We normalize so that (in homogeneous coordinates)

$$P_{-7} = \begin{pmatrix} 0 \\ x_0x_1 \\ 1 \end{pmatrix}, \quad P_{-3} = \begin{pmatrix} 0 \\ 0 \\ 1 \end{pmatrix}, \quad P_1 = \begin{pmatrix} 1 \\ 0 \\ 0 \end{pmatrix}, \quad P_5 = \begin{pmatrix} 1 \\ 1 \\ 0 \end{pmatrix} \tag{7.1}$$

We define polynomials O_a^b for $a \leq b+2$ odd integers, in the following recursive way. First, for all b we set $O_b^b = O_{b-2}^b = 1$ and $O_{b+2}^b = 0$. Next, we define

$$O_a^b = \begin{pmatrix} 1 \\ -x_{b-2} \\ x_{b-4}x_{b-3}x_{b-2} \end{pmatrix} \cdot \begin{pmatrix} O_a^{b-2} \\ O_a^{b-4} \\ O_a^{b-6} \end{pmatrix}, \qquad a = b-4, b-6, \ldots \tag{7.2}$$

These polynomials are quite close to the pentagram invariants O_k defined in §5. The rest of the points of the polygon are given by

$$P_{9+2k} = \begin{pmatrix} O_{-1}^{3+k} \\ O_{+1}^{3+k} \\ O_{+3}^{3+k} \end{pmatrix}, \qquad k = 0, 2, 4, \ldots \tag{7.3}$$

The polygon P with vertices $P_{-7}, P_{-3}, P_1, \ldots$ has flag invariants $x_0, x_1, x_2, \ldots$.

If we just know the 8 points $P_{-7}, \ldots, P_{21}$ then we can compute the invariants $x_0, \ldots, x_7$. We have listed formulas for the first four of these points above, and now we list formulas for the next four of these points. To simplify the expressions, we introduce the monomials

$$X_k = x_{k-1}x_kx_{k+1}. \tag{7.4}$$

The points are then given by

$$P_9 = \begin{pmatrix} 1 - x_1 \\ 1 \\ 1 \end{pmatrix},$$

$$P_{13} = \begin{pmatrix} 1 - x_1 - x_3 + X_2 \\ 1 - x_3 \\ 1 \end{pmatrix},$$

$$P_{17} = \begin{pmatrix} 1 - x_1 - x_3 - x_5 + X_2 + X_4 + x_1x_5 \\ 1 - x_3 - x_5 + X_4 \\ 1 - x_5 \end{pmatrix},$$

$$P_{21} = \begin{pmatrix} 1 - x_1 - x_3 - x_5 - x_7 + X_2 + X_4 + X_6 + x_1x_5 + x_3x_7 + x_1x_7 - X_2x_7 - x_1X_6 \\ 1 - x_3 - x_5 - x_7 + X_4 + X_6 + x_3x_7 \\ 1 - x_5 - x_7 + X_6 \end{pmatrix} \tag{7.5}$$

7.2. The Dual Map

There is one more piece of structure we want to explain before we derive the formula for the projective heat map. We want to explain what we call the *dual projective heat map*, H^*. The map H^* is essentially obtained from the projective heat map H by dualizing the construction. In this section we explain what that means. The reason we consider both H and H^* in general is that the formulas for the one map are intimately bound up with the formulas of the other.

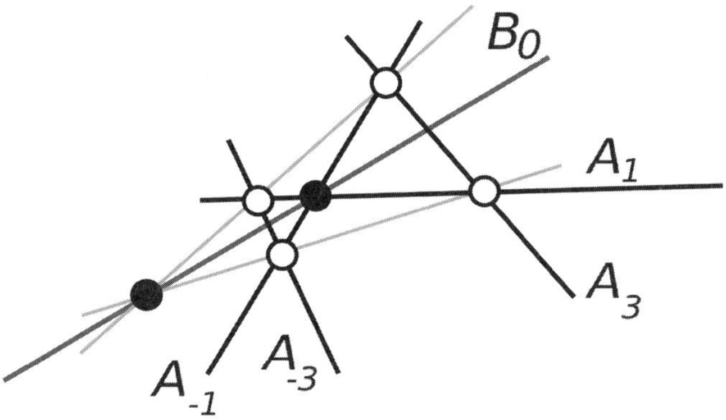

FIGURE 7.1. The construction of B_0 from A_{-3}, A_{-1}, A_1, A_3

Given 4 lines A_{-3}, A_{-1}, A_1, A_3, we construct the new line B_0 shown in Figure 7.1.

Equation (3.1) also gives an equation for B_0. It is the same equation as for the projective midpoint, except that the roles of points and lines are reversed.

Starting with a polygon P described in terms of its lines $A_1, A_3, \ldots$, we define $H^*(P)$ to be the polygon whose lines are $B_0, B_2, \ldots$. On the level of unlabeled polygons, we have

$$H^* = \Delta^* \circ H \circ \Delta^*. \tag{7.6}$$

for any polarity Δ. See §3.3 for the definition of a polarity. Here $\Delta^*(P)$ is the polygon whose lines are the images of the vertices of P under Δ.

When P is a pentagon, $\Delta^*(P) = P$ up to projective equivalence. For this reason, H and H^* induce the same map on $\mathcal{P}_5$. This makes the map especially canonical in this case. In general H and H^* induce different maps. We will study the interaction between H and H^* below.

7.3. Formulas for the Projective Heat Map

As with the other maps we have considered, we must break symmetry to get a labeling scheme for the action of H or H^* on $\mathcal{P}_N$. However, there is a canonical choice for $\mathcal{P}_5$. This is shown in Figure 7.2.

The maps H^2 and HH^* and H^*H and $(H^*)^2$ all have canonical labeling schemes when they act on $\mathcal{P}_N$. Figure 7.3 shows the canonical labeling scheme for the action of H^2. The labeling schemes for the other maps are similar, and we will discuss them below.

Choose a polarity Δ and define

$$\Omega = \Delta^* \circ H, \tag{7.7}$$

In other words, $\Omega(P)$ is the polygon whose edges are the edges of $\Delta(H(P))$. The action on $\mathcal{P}_N$ is independent of the choice of polarity. There is a canonical labeling scheme for Ω. The kth edge of $H(P)$ has its endpoints in the projective midpoints lying on the two edges incident to the kth vertex of P. The kth vertex of $\Omega(P)$ is the image of the kth edge of $H(P)$ under Δ^*.

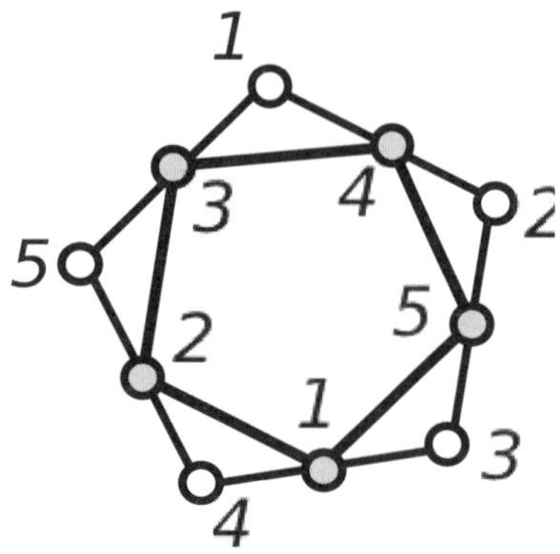

FIGURE 7.2. The canonical labeling scheme for H acting on $\mathcal{P}_5$

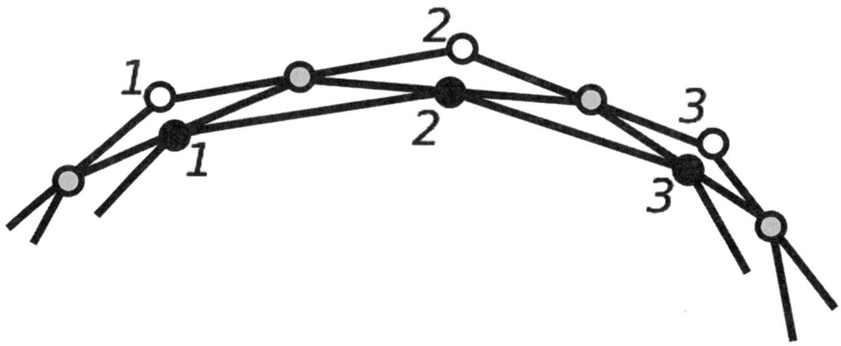

FIGURE 7.3. The canonical labeling scheme for H^2

We also have the left and right shift maps L and R, which respectively shift the indices one unit to the left or to the right. These maps act on $\boldsymbol{R}^{2n}$ and have order $2n$. We have the following relations:

- Ω^2 gives the formula for H^*H with its canonical labeling scheme.
- $L\Omega = H$ relative to some labeling scheme. The same goes for $R\Omega$.
- $\Omega L = H^*$ relative to some labeling scheme. The same goes for ΩR.
- $L\Omega R\Omega L = R\Omega L\Omega R = H^2$ relative to the canonical labeling scheme.
- $\Omega R\Omega L = \Omega L\Omega R = (H^*)^2$ relative to the canonical labeling scheme.
- $\Omega = H = H^*$ on $\mathcal{P}_5$ relative to the canonical labeling schemes.

To compute the formula for Ω, we write out the first 8 points of P, as given in equations (7.1) and (7.5). We then construct the first 5 points of Q using the straight line construction. Finally, we compute the only 2 flag invariants of Q we have enough information to compute. We do the calculation in Mathematica. (The reader who downloads our computer code can see the Mathematica code.) Given the formula for Ω, we readily get the formulas for the other maps listed above.

These calculations are done behind the scenes so to speak, and now we present the result. Suppose that P has projective flag invariants $x_0, x_1, x_2, \ldots$ and $Q = H(P)$

has flag invariants $y_0, y_1, y_2, \ldots$. Define

$$A_k = 2 + x_{k-3/2} + x_{k+3/2}, \tag{7.8}$$

$$B_k^{\pm} = A_{k\mp 1} - x_{k\pm 1/2}A_{k\pm 1} + x_{k\pm 5/2}A_{k\mp 1}, \tag{7.9}$$

$$\Omega(x_1, \ldots, x_{2n}) = (y_1, \ldots, y_{2n}), \tag{7.10}$$

where

$$\frac{y_{2k+0}}{x_{2k+1}} = \frac{A_{2k-5/2}B^-_{2k+1/2}}{B^+_{2k-3/2}B^+_{2k+1/2}}, \qquad \frac{y_{2k+1}}{x_{2k+0}} = \frac{A_{2k+5/2}B^+_{2k+1/2}}{B^-_{2k+3/2}B^-_{2k+1/2}}. \tag{7.11}$$

Of course, one can solve for y_{2k+0} and y_{2k+1}.

It seems worth unpacking this formula a bit. Concretely, we have

$$y_5 = \frac{x_6(2 + x_1 + x_4)(-2 - 2x_3 + x_5 + x_5x_6 - x_8 - x_3x_8)}{(2 + x_1 - x_4 - x_3x_4 + 2x_6 + x_1x_6)(-2 - x_3 + x_6 + x_5x_6 - 2x_8 - x_3x_8)}.$$

The formula for y_6 is obtained from the formula for y_5 by replacing each index k by $11 - k$. The formula for the remaining y variables is obtained by cyclically shifting the formulas for y_5 and y_6 by suitably chosen even amounts.

7.4. The Case of Pentagons

In the special case $N = 5$ we can use the Gauss group, defined in §3.6, to express everything in terms of the first two coordinates, which we call $x = x_1$ and $y = x_2$. The general formula simplifies somewhat. We then take the first two coordinates, and this gives us the rational map H in equation (1.1).

The first thing we can do with the formula is compute the fixed points. Let ϕ denote the golden ratio. The first fixed point of H is the regular class $r = (\phi^{-1}, \phi^{-1})$, the point corresponding to the regular class. We compute the differential

$$dH_r = \begin{bmatrix} 2\phi^{-5} & 0 \\ 0 & 2\phi^{-5} \end{bmatrix}, \tag{7.12}$$

Thus, the point representing the regular class is an attracting fixed point, with multiplier $2\phi^{-5}$.

The point $(-\phi, -\phi)$ corresponds to the star regular class. It is a repelling fixed point, and

$$dH_{(-\phi,-\phi)} = \begin{bmatrix} \frac{-\phi^5}{2} & 0 \\ 0 & \frac{-\phi^5}{2} \end{bmatrix}. \tag{7.13}$$

The product of the multipliers at (ϕ^{-1}, ϕ^{-1}) and $(-\phi, -\phi)$ is -1.

H has 5 additional fixed points. One of the points is $(-3, -3)$, and the other ones are the orbit of $(-3, -3)$ under the Gauss group:

$$(-3, -1/2), \qquad (-1/2, -8), \qquad (-8, -1/2), \qquad (-1/2, -3). \tag{7.14}$$

These fixed points are all repelling. The eigenvalues of the linear differential at each of these points are $-22/3$ and 6. The corresponding eigenvectors are $(-1, 1)$ and $(1, 1)$. This structure will be important for the discussion in §16.4.

The 5 fixed points just mentioned correspond to consecutive flag invariants of the star convex isosceles pentagon with vertices.

$$(-1, -1), \qquad (0, 1), \qquad (0, -1), \qquad (-1, 1), \qquad (1, 0).$$

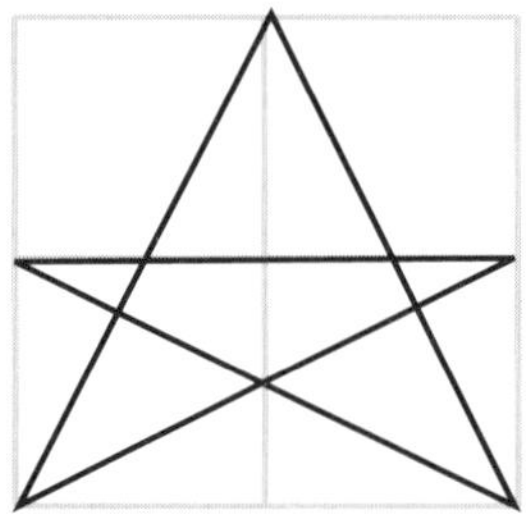

FIGURE 7.4. A particular star isosceles pentagon

We mentioned in the introduction that eventually we will conjugate H by the map $\boldsymbol{B}$ from equation (1.2). Here we explain the motivation for that. We have

$$\boldsymbol{B}(-\phi,-\phi)=(0,0), \qquad \boldsymbol{B}(-3,-3)=(1,1), \qquad B(1/\phi,1/\phi)=(\infty,\infty). \tag{7.15}$$

Again, ϕ is the golden ratio. So, when we change coordinates, the origin represents the star-regular class and (∞,∞) represents the regular class. This turns out to be a very convenient normalization for most of the analysis. The map $\boldsymbol{H}=\boldsymbol{B}H\boldsymbol{B}^{-1}$ is an algebraic monster – see our computer program for a display of the equation – but we will deal with it in a computer assisted way.

7.5. Some Speculation

The maps H and H^* are not invertible but we can consider the semigroup $\mathcal{H}$ generated by H and H^*. Here is a conjecture which is based mostly on experiments done in the case $N=6$ and $N=7$.

CONJECTURE 7.1. *Let $G\in\mathcal{H}$ be any nontrivial element which does not have the form $(HH^*)^k$ or $(H^*H)^k$ for $k=1,2,3,\ldots$ Then almost every point $P\in\mathcal{P}_N$ is such that $\{G^n(P)\}$ converges to the projectively regular class.*

Now we describe what happens for the two maps H^*H and HH^* in case $N=6$ and $N=7$. Let c be the string (a,b,b,a) and let d be the string (b,a,a,b). The string (c,c) stands for (a,b,b,a,a,b,b,a), and so on. A calculation shows that

$$\Omega(c,\ldots,c)=(d,\ldots,d). \tag{7.16}$$

Hence $\Omega^2=H^*H$ fixes any point in $\mathcal{P}_N$ having this form. Similarly, HH^* fixes points of the form $(a,a,b,b,a,a,b,b\ldots)$ when such strings have length which is a multiple of 4. When $N=6$ these two kinds of points correspond to the projective classes of hexagons having 6-fold dihedral symmetry. Figure 7.5 shows the shapes of the hexagons corresponding to these two kinds of points.

Let's call these two kinds of hexagons type 1 and type 2. Here is what one sees experimentally.

CONJECTURE 7.2. *Let $F=H^*H$ (respectively $F=HH^*$). Almost every $P\in\mathcal{P}_6$, and every $P\in\mathcal{C}_6$, has the property that $\{F^n(P)\}$ converges to the class of a convex hexagon of type 1 (respectively type 2).*

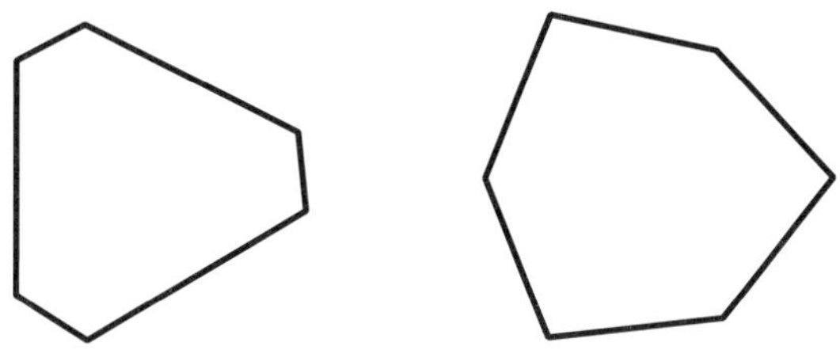

FIGURE 7.5. Symmetric hexagons fixed by H^*H and HH^* respectively

The situation on $\mathcal{P}_7$ is strange. Recall that $f(P)$ is the product of the flag invariants of P, as discussed in §5.3. Letting P_0 be the regular class, we have

$$f(P_0) = \Big(2\sin(\pi/14)\Big)^{14} \approx \Big(.445\Big)^{14}. \tag{7.17}$$

Let

$$\eta = \Big(\frac{7}{18}\Big)^{14} \approx \Big(.38888\Big)^{14}. \tag{7.18}$$

There is nothing significant about 7/18 in this definition, except that it is a rational approximation to a slightly smaller number which starts .38887. The significant thing for us is that $\eta < f(P_0)$.

CONJECTURE 7.3. *Let $F = H^*H$ or $F = HH^*$. Almost every $P \in \mathcal{P}_7$, and every $P \in \mathcal{C}_7$, has the property that $\{F^n(P)\}$ converges to some $Q \in \mathcal{C}_N$ with $f(Q) < \eta$.*

The polygon Q depends on the initial choice on P. It looks like $f(Q)$ also depends on P, but only in a mild way. It almost always happens that $f(Q) =$ .38887... I'm not sure whether the value is always the same.

For fun, I thought about the larger semigroup $\langle H, H^*, T, T^*\rangle$. Here T is the pentagram map and $T^* = T^{-1}$ is the conjugation of T by a duality. At first there seems to be no connection between H and T, but then there are some curious results which seem to suggest that the maps are not entirely unrelated.

CONJECTURE 7.4. *Let F be any word that just involves T and H. Then for almost all $P \in \mathcal{P}_N$, the sequence $\{F^n(P)\}$ converges to the regular class.*

On the other hand, the element $F = H^2T^*$ seems to have the following structure: For $N \geq 11$ the map seems to obey Conjecture 7.4. However, for $N \geq 12$ it seems that the space $\mathcal{P}_N$ has an attracting submanifold A_N so that almost all $P \in \mathcal{P}_N$ have the property that $\{F^n(P)\}$ converges to A_N. The restriction of F to A_N seems to be torus motion, though I am not sure.

Here is another example of strange behavior. For $N = 7$, all the words

$$(HH^*)^k HT^2, \qquad k = 1, 2, 3, \ldots$$

seem to obey a version of Conjecture 7.3 but with different constants. Most of the other words in the semigroup generated by H,H^* and T seem to obey Conjecture 7.4.

My computer program has a subdirectory with an auxiliary program that lets the user do experiments with the action of the semigroup on N-gons. So, the interested reader can play around and draw his or her own conclusions.

CHAPTER 8

Topological Degree of the Map

8.1. Overview

The purpose of this chapter is to prove Theorem 1.2. We remind the reader of this result. We have the rational map

$$H(x,y) = (x', y'),$$

$$x' = \frac{(xy^2+2xy-3)\,(x^2y^2-6xy-x+6)}{(xy^2+4xy+x-y-5)\,(x^2y^2-6xy-y+6)}$$

$$y' = \frac{(x^2y+2xy-3)\,(x^2y^2-6xy-y+6)}{(x^2y+4xy-x+y-5)\,(x^2y^2-6xy-x+6)} \tag{8.1}$$

We will prove that a generic point $p \in \boldsymbol{C}^2$ has the property that $H^{-1}(p)$ consists of 6 points. Here, *generic* means that this statement is true for a set $U \subset \boldsymbol{C}^2$ such that $\boldsymbol{C}^2 - U$ is the solution set of a nontrivial polynomial. Such sets are called *Zariski open*. They are open and dense, and have full measure.

In general, there is some constant M such that $H^{-1}(p)$ has M pre-images for all p in a Zariski open set. We just want to prove that $M = 6$. We will show that $M \geq 6$ and that $M \leq 6$. The lower bound is a direct calculation and the upper bound is an application of Bezout's Theorem.

8.2. The Lower Bound

Here we prove that $M \geq 6$.

Let

$$\phi = (1+\sqrt{5})/2 \tag{8.2}$$

be the golden ratio. Let $r = (\phi^{-1}, \phi^{-1})$. In our coordinates, r represents the regular class. By symmetry $H(r) = r$. Next, define

$$q_1 = q_2 = \phi - 2, \qquad q_3 = q_5 = \phi, \qquad q_4 = 3\phi - 4. \tag{8.3}$$

The funny ordering of the points is deliberate. We check that $H(q_i, q_{i+1}) = r$ for $i = 0, 1, 2, 3, 4$, with indices taken mod 5.

Hence $H^{-1}(r)$ has at least 6 points. Next, we check that that the Jacobian J_H does not vanish at any of these 6 points. Hence, there is an open set in $\boldsymbol{C}^2$ such that every $q \in U$ has at least 6 pre-images.

8.3. The Upper Bound

The coordinate change

$$x = Y/X, \qquad y \to X \tag{8.4}$$

is a birational coordinate change. So, it suffices to prove that equation

$$H(Y/X, X) = (a_1, a_2) \tag{8.5}$$

generically has at most 6 solutions in $\boldsymbol{C}^2$. When we expand out equation (8.5) and then homogenize we get

$$\left(\frac{P_1 + a_1 Q_1}{R_1}, \frac{P_2 + a_2 Q_2}{R_2}\right) = (0, 0), \tag{8.6}$$

$$P_1 = -(XY + 2YZ - 3Z^2)(XY^2 - 6XYZ + 6XZ^2 - YZ^2).$$
$$Q_1 = -(Y^2 + XZ + 6YZ - 6Z^2)(X^2Y - X^2Z + 4XYZ - 5XZ^2 + YZ^2).$$
$$R_1 = (-5XZ^2 - X^2Z + YZ^2 + 4XYZ + X^2Y)(6Z^2 - XZ - 6YZ + Y^2).$$
$$P_2 = -X(2XY + Y^2 - 3XZ)(-Y^2 + XZ + 6YZ - 6Z^2).$$
$$Q_2 = -(X^2 + 4XY + Y^2 - 5XZ - YZ)(XY^2 - 6XYZ + 6XZ^2 - YZ^2).$$
$$R_2 = (-5XZ + XZ + YZ + 4X + Y^2)(6XZ^2 - YZ^2 - 6XYZ + XY^2).$$

Before applying Bezout's Theorem, we need to take care of a technical point. In the technical point, we will make use of the Resultant of a polynomial, but only in a theoretical sense. All we need to know is that the resultant of two polynomials is a polynomial expression in their coefficients which vanishes if and only if the polynomials have a common factor.

LEMMA 8.1. *For a generic choice of $a_1, a_2 \in \boldsymbol{C}^2$ the polynomials $P_1 + a_1 Q_1$ and $p_2 + a_2 Q_2$ have no common factors.*

Proof: Suppose that this result is false. Then, after we dehomogenize by setting $Z = 1$, the set of points where the polynomials have a common factor is not Zariski open. But then the resultant of these polynomials vanishes everywhere. But then, by continuity, the polynomials have a common factor everywhere, including at the values $a = -1$ and $b = -1$. But one can check in Mathematica (or by trial and error) that both $P_1 - Q_1$ and $P_2 - Q_2$ are irreducible and not scalar multiples of each other. □

Set $f_j = P_j + a_j Q_j$, with the understanding that these functions depend on a_1 and a_2. Inspecting our equations, we see that for the generic choice of (a_1, a_2), we have

$$\deg(f_j) = 5, \qquad j = 1, 2. \tag{8.7}$$

Hence $V(f_1, f_2)$ has 25 points (properly counted), by Bezout's Theorem.

Define

$$\psi = \frac{\sqrt{13} + 1}{2} \tag{8.8}$$

We find the following points in $V(f_1, f_2)$, independent of the choice of a_1, a_2:

- $[1, 0, 0]$ with intersection number 4.
- $[1, 1, 1]$ with intersection number 4.
- $[0, 1, 0]$ with intersection number 2.

- $[0,0,1]$ with intersection number 2.
- $[-2,4,1]$.
- $[+\psi+0,-\psi+3,1]$.
- $[+\psi+0,+\psi+3,1]$.
- $[+\psi-2,-\psi+3,1]$.
- $[-\psi+1,+\psi+2,1]$.
- $[-\psi+1,-\psi+4,1]$.
- $[-\psi-1,+\psi+2,1]$.

We check that these points all lie in $V(P_1,Q_1,R_1,P_2,Q_2,R_2)$. All the component polynomials vanish. Hence, these points all belong to $V(f_1,f_2)$, independent of a_1 and a_2. They account for 19 of the solutions. Since R_1 and R_2 vanish on these points, the map H is not well defined on the corresponding points. To be sure, we check this directly. For instance, the point $[-\psi,\psi+3,1]$ corresponds to the point

$$\left(\frac{\psi+3}{-\psi},-\psi\right)=\left(\frac{-3-\sqrt{13}}{2},\frac{1-\sqrt{13}}{2}\right)$$

and this is indeed an indeterminate point for H.

Since 19 of the 25 solutions in $V(f_1,f_2)$ do not correspond to pre-images of H in $\boldsymbol{C}^2$, there can be at most 6 pre-images of a generic point. In outline, this completes the proof.

However, we are not quite done. We have to verify that the first 4 points listed above have the advertised intersection numbers. We treat these points in turn. Actually, we will be a bit lazy and we will just show that each of the points has at least the advertised intersection number. Bezout's theorem then tells us that the we have equality in all cases.

Case 1: For $[1,0,0]$ we compute the intersection number in the affine patch $\{X\neq 0\}$. Let P_j^* etc. be the polynomials we get by setting $X=1$. We can see directly that each of the factors of P_j^* and Q_j^* vanishes at $(0,0)$. Hence $(0,0)$ vanishes to at least second order for each P_j^* and Q_j^*. Hence $(0,0)$ vanishes to at least second order for each f_j^*. But then $I_{(0,0)}(f_1^*,f_2^*)\geq 2\times 2=4$ on $[1,0,0]$. Hence $[1,0,0]$ has intersection number at least 4.

Case 2: For $[1,1,1]$ we make the change of variables $X=1+X'$ and $Y=1+Y'$ and then repeat exactly the same argument for the polynomials in the X' and Y' variables. It works exactly the same way. Hence $[1,1,1]$ has intersection number at least 4.

Case 3: For $[0,1,0]$ we consider the picture in the affine patch $\{Y=1\}$. We check that P_1^* and Q_1^* vanish to second order at $(0,0)$, and P_2^* and Q_2^* both vanish (to first order). Hence $[0,1,0]$ has intersection number at least 2.

Case 4: For $[0,0,1]$ we consider the picture in the affine patch $\{Z=1\}$. We check that P_2^* and Q_2^* vanish to second order at $(0,0)$, and P_1^* and Q_1^* both vanish (to first order). Hence $[0,1,0]$ has intersection number at least 2.

Everything works out as claimed. This completes the proof of Theorem 1.2.

CHAPTER 9

The Convex Case

9.1. Flag Invariants of Convex Pentagons

Recall that $\mathcal{C}$ is the set of projective classes of convex pentagons. The purpose of this chapter is to prove Theorem 1.3. This result says that for every convex class $P \in \mathcal{C}$ we have $f(H(P)) \geq f(P)$, with equality iff P is the regular class. Here f is the first pentagram invariant, defined in §5.3. When P is convex, all the flag invariants lie in $(0,1)$. Since $f(P)$ is the product of the flag invariants for P, we have $f(P) \in (0,1)$ when P is convex.

Our flag coordinates identify $\mathcal{C}$ with a subset of $(0,1)^2$. In fact, this identification is a diffeomorphism.

LEMMA 9.1. *Taking two consecutive flag invariants of a pentagon gives a diffeomorphism from $\mathcal{C}$ onto the open unit square $(0,1)^2$. In particular, every $(x,y) \in (0,1)^2$ arises as a list of 2 consecutive flag invariants of a convex pentagon.*

Proof: When we normalize our pentagon so that it has points p_1, p_2, p_3, p_4, p_5 as in the the proof of Lemma 3.2, we can identify $\mathcal{C}$ with the space of choices of p_5 in the domain $D = (0,1) \times (1, \infty)$. The map

$$(x,y) \to \left(\frac{x-1}{x-y}, \frac{x}{x+y-1}\right),$$

which comes from computing the consecutive flag invariants associated to p_5, is a diffeomorphism between D and $(0,1)^2$. The inverse map is given by

$$(x,y) \to \left(\frac{xy-y}{2xy-x-y}, \frac{xy-1}{2xy-x-y}\right).$$

Were we to choose different consecutive flag invariants, we would get the same result, by projective invariance and by the self-dual nature of pentagons. □

COROLLARY 9.2. A pentagon is convex if and only if any 2 consecutive flag invariants of the pentagon lie in $(0,1)^2$.

Proof: We have already explained that a pentagon has flag coordinates in $(0,1)^2$ provided it is convex. If a nonconvex pentagon had flag coordinates in $(0,1)^2$ then its flag coordinates would coincide with those of some convex pentagon, by the previous result. But this is a contradiction, because the flag coordinates determine the pentagon. □

9.2. The Gauss Group Acting on the Unit Square

We now study how the Gauss recurrence G, defined in §3.8, acts on $(0,1)^2$.

LEMMA 9.3. *The open square* $(0,\phi^{-1})^2$ *contains a fundamental domain for the action of* G *on* $(0,1)^2 - (\phi^{-1},\phi^{-1})$.

Proof: Let $r = (\phi^{-1},\phi^{-1})$. Note that $G(r) = r$. Let s be the line segment joining r to $(1,1)$. Recall that $R(x,y) = (y,z)$. Note that R is just a reflection in the line extending s. We compute that $G^3(s) = RG_2(s)$. From the formula

$$G^2(x,x) = (\frac{1}{1+x}, 1-x^2). \tag{9.1}$$

We see that $\gamma = G^2(s)$ is a curve which connects r to $(1/2,0)$ and remains inside $(0,\phi^{-1})^2$ except at r.

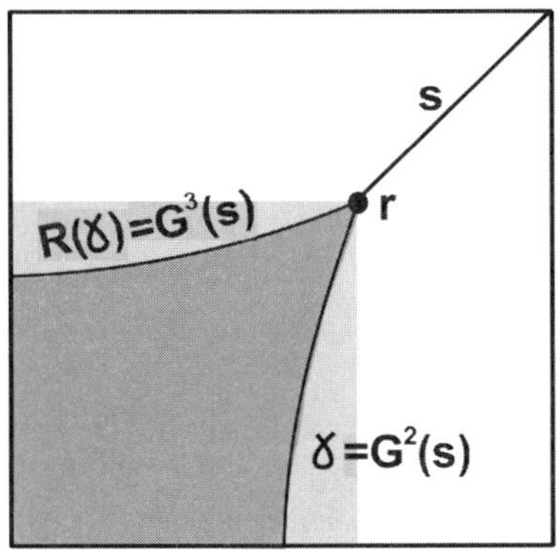

FIGURE 9.1. Fundamental domain for the action of G

But then a fundamental domain for the action of Γ on $(0,1)^2 - \{r\}$ is contained in open region bounded by the coordinate axes and $G^2(\gamma)$ and $G^3(\gamma)$. This region is contained in $(0,\phi^{-1})^2$. $\square$

9.3. A Positivity Criterion

We consider real polynomials in the variables $x_1,\ldots,x_k$. For our purposes in this monograph, we take $k=2$. However, the proof in the lemma below goes by induction on k, and there is no harm in presenting the general case. The criterion presented here is very useful for proving Theorem 1.3 but not very useful in general. We will discuss a more powerful criterion in §11.

Given a multi-index $I = (i_1,\ldots,i_k) \in (\boldsymbol{N}\cup\{0\})^k$ we let

$$x^I = x_1^{i_1}\ldots x_k^{i_k}. \tag{9.2}$$

Any polynomial $F \in \boldsymbol{R}[x_1,\ldots,x_k]$ can be written succinctly as

$$F = \sum a_I X^I, \qquad a_I \in \boldsymbol{R}. \tag{9.3}$$

If $I' = (i'_1, \ldots, i'_k)$ we write $I' \leq I$ if $i'_j \leq i_j$ for all $j = 1, \ldots, k$. We call F *very weak positive dominant* (VWPD) if

$$A_I := \sum_{I' \leq I} a_{I'} \geq 0 \qquad \forall I, \tag{9.4}$$

We call F *weak positive dominant* if F is WVPD and the total sum, $\sum a_I$, is positive. We call F *positive dominant* if $A_I > 0$ for all I.

LEMMA 9.4. *The following is true.*

1. *If P is VWPD, then $P \geq 0$ on $[0,1]^k$.*
2. *If P is WPD, then $P > 0$ on $(0,1)^k$.*
3. *If P is PD, then $P > 0$ on $[0,1]^k$.*

Proof: The proof goes by double induction and breaks into $6 = 3 \times 2$ cases. We first prove each of the above statements for $k = 1$, by induction on the degree. Then we prove the general case of each of the above statements by induction on k.

Case 1A: We first prove Statement 1 when $k = 1$. The case $\deg(P) = 0$ follows from the fact that $a_0 = A_0 \geq 0$. Let $x \in [0,1]$. We have

$$P(x) = a_0 + a_1 x + x_2 x^2 + \cdots + a_n x^n \geq$$

$$a_0 x + a_1 x + a_2 x^2 + \cdots + a_n x^n =$$

$$x(A_1 + a_2 x + a_3 x^2 + \cdots a_n x^{n-1}) = xQ(x) \geq 0. \tag{9.5}$$

The final inequality is induction on the degree: Q is VWPD and has degree $n-1$.

Case 2A: Now we prove Statement 2 when $k = 1$. In this case we have $P > 0$ when $\deg(P) = 0$. Equation (9.5) holds word for word, except for the following improvement: When $x \in (0,1)$ we get $xQ(x) > 0$ because Q is WPD of degree $n-1$.

Case 3A: Now we prove Statement 3 when $k = 1$. If P is PD then P is also WPD, so the previous argument shows that $P > 0$ on $(0,1)$. When $x = 0$ we have $P(0) = a_0 > 0$. Finally, referring to equation (9.5), we have $P(1) = Q(1) > 0$ because Q is PD of degree $n-1$.

Case 1B: Now we prove the general case of Statement 1. Suppose the the coefficients of P are $\{a_I\}$. We write

$$P = f_0 + f_1 x_k + \cdots + f_m x_k^m, \qquad f_j \in \boldsymbol{R}[x_1, \ldots, x_{k-1}]. \tag{9.6}$$

Let $P_j = f_0 + \cdots + f_j$. A typical coefficient in P_j has the form

$$b_J = \sum_{i=1}^{j} a_{Ji}, \tag{9.7}$$

where J is a multi-index of length $k-1$ and Ji is the multi-index of length k obtained by appending i to J. From equation (9.7) and the definition of VWPD, the fact that P is WVPD implies that P_j is WVPD for all j.

By induction on k, we get $P_j \geq 0$ on $[0,1]^{k-1}$. But now, if we hold $x_1, \ldots, x_{k-1}$ fixed and let $t = x_k$ vary, the polynomial $g(t) = P(x_1, \ldots, x_{k-1}, t)$ is VWPD. The point here is that the sum of the first j coefficients of $g(t)$ is precisely $P_j(x_1, \ldots, x_{k-1})$. Hence, by the single variable case, $g \geq 0$ on $[0,1]$. Since this is true for any choices of the other variables, we see that $P \geq 0$ on $[0,1]^k$.

Case 2B: Now we prove the general case of Statement 2. Since P is also VWPD, the same argument as in Case 1A shows that the single variable polynomial g is VWPD for any choice of variables $x_1, \ldots, x_{k-1} \in [0,1]^{k-1}$. Now we use the extra information. When P is WPD, the polynomial P_k is also WPD, because the total sum of the coefficients of P_k is the same as the total sum of the coefficients of P. But then $P_k(x_1, \ldots, x_{k-1}) > 0$ when $x_1, \ldots, x_{k-1} \in (0,1)^{k-1}$. But this nonzero value is precisely the value of the sum of the coefficients of g for this choice of variables. Hence g is WPD when $(x_1, \ldots, x_{k-1}) \in (0,1)^{k-1}$. Hence $g(t) > 0$ for $t \in (0,1)$ and all $(x_1, \ldots, x_{k-1}) \in (0,1)^{k-1}$. This is what we wanted to prove.

Case 3B: The proof is the same as Case 2B except that this time P_j is PD for all j. Hence $g(t)$ is PD for any choice of $x_1, \ldots, x_{k-1} \in [0,1]^{k-1}$. Hence $g(t) > 0$ for all $t \in [0,1]$ and $(x_1, \ldots, x_{k-1}) \in [0,1]^{k-1}$. This is what we wanted to prove. □

Remark: In the argument below we only use the WPD case of the above lemma. However, in later chapters we will also deal with the other cases. It seemed worth treating all three cases at once.

9.4. The End of the Proof

Let f be the first pentagram invariant, written in flag coordinates. If the first two flag invariants of P are x and y, then the Gauss Recurrence from equation (3.16) gives the remaining flag invariants as $(1-x)/(1-xy)$ and $1-xy$ and $(1-y)/(1-xy)$. Hence

$$f(P) = \frac{xy(1-x)(1-y)}{1-xy}. \tag{9.8}$$

Using Mathematica [**W**], we compute that

$$f(H(P)) = \frac{(-1+xy)(-3-x+xy)(-3-y+xy)}{(4+x+y)(-5-x+y+4xy+x^2y)(-5+x-y+4xy+xy^2)}$$

$$\times \frac{(-4+x+y+2xy)(-3+2xy+x^2y)(-3+2xy+xy^2)}{(6-x-6xy+x^2y^2)(6-y-6xy+x^2y^2)} \tag{9.9}$$

We have split things up this way simply because the equation is too long to fit on one line. We compute that

$$f(H(x,y)) - f(x,y) = \frac{\Upsilon(x,y)}{D(x,y)} \tag{9.10}$$

$D(x,y)$ is some polynomial whose formula we do not care about and

$$\begin{gathered}\Upsilon(x,y) = -x^{10}y^9 - 3x^{10}y^8 + 3x^{10}y^7 + x^{10}y^6 \\ -x^9y^{10} - 10x^9y^9 - 4x^9y^8 + 66x^9y^7 - 34x^9y^6 - 16x^9y^5 - x^9y^4 \\ -3x^8y^{10} - 4x^8y^9 + 124x^8y^8 + 227x^8y^7 - 537x^8y^6 + 107x^8y^5 + 79x^8y^4 + 7x^8y^3 \\ +3x^7y^{10} + 66x^7y^9 + 227x^7y^8 - 504x^7y^7 - 1761x^7y^6 + 2132x^7y^5 + 16x^7y^4 - 140x^7y^3 - 12x^7y^2 \\ +x^6y^{10} - 34x^6y^9 - 537x^6y^8 - 1761x^6y^7 + 814x^6y^6 + 6231x^6y^5 - 4496x^6y^4 - 481x^6y^3 + 95x^6y^2 + 6x^6y \\ -16x^5y^9 + 107x^5y^8 + 2132x^5y^7 + 6231x^5y^6 - 1564x^5y^5 - 12565x^5y^4 + 5114x^5y^3 + 660x^5y^2 - 18x^5y \\ -x^4y^9 + 79x^4y^8 + 16x^4y^7 - 4496x^4y^6 - 12565x^4y^5 + 6034x^4y^4 + 15227x^4y^3 - 2941x^4y^2 - 273x^4y \\ +7x^3y^8 - 140x^3y^7 - 481x^3y^6 + 5114x^3y^5 + 15227x^3y^4 - 12842x^3y^3 - 10404x^3y^2 + 684x^3y \\ -12x^2y^7 + 95x^2y^6 + 660x^2y^5 - 2941x^2y^4 - 10404x^2y^3 + 12650x^2y^2 + 3057x^2y - 27x^2 \\ +6xy^6 - 18xy^5 - 273xy^4 + 684xy^3 + 3057xy^2 - 5076xy + 27x \\ -27y^2 + 27y + 324.\end{gathered}$$

LEMMA 9.5. *$D(x,y) > 0$ provided that $\Upsilon(x,y) > 0$ on $(0,1)^2 - (\phi^{-1},\phi^{-1})$.*

Proof: We have Since P and $H(P)$ are both convex, both $f(P)$ and $f(H(P))$ lie in $(0,1)$. Hence

$$|f(H(P)) - f(P)| \le 1,$$

so $D(x,y)$ can only vanish when Υ vanishes. So, if Υ only vanishes at (ϕ^{-1},ϕ^{-1}), then either Υ and D always have the same sign, or always have opposite signs. We check at a single point that they have the same sign. $\square$

Now we are going to use our positivity criterion to deal with Υ. The criterion doesn't apply right away. We have to fool around with the polynomial a bit. This is why we proved Lemma 9.3.

Define

$$\Upsilon_1(x,y) = \Upsilon\left(\frac{1-x}{\phi}, \frac{1-y}{\phi}\right). \tag{9.11}$$

Note that $\Upsilon_1 > 0$ on $(0,1)^2$ if and only if $\Upsilon > 0$ on $(0,\phi^{-1})^2$. But then, by Lemma 9.3 and symmetry, we see that $\Upsilon > 0$ on $(0,\phi^{-1})^2$ if and only if $\Upsilon > 0$ on $(0,1)^2 - \{r\}$, where $r = (\phi^{-1},\phi^{-1})$. So, to finish the proof, we just have to prove the following lemma.

LEMMA 9.6. *$\Upsilon_1 > 0$ on $(0,1)^2$.*

Proof: Listing out the lowest order nontrivial terms, we have

$$\Upsilon_1(x,y) = a_{20}x^2 + a_{11}xy + a_{02}y^2 + \ldots,$$

$$a_{20} = a_{02} = 320 - 120\sqrt{5}, \qquad a_{11} = 460 - 220\sqrt{5}. \tag{9.12}$$

Since $a_{00} = a_{10} = a_{01} = 0$ and $a_{11} < 0$, we have $A_{11} < 0$. This means that Υ_1 is not WPD.

Define

$$\Upsilon_2(x,y) = \Upsilon_1(x,y) + \frac{a_{11}}{2}(x-y)^2. \tag{9.13}$$

We have $\Upsilon_2 \le \Upsilon_1$. We check by direct calculation that Υ_2 is WPD. With respect to Υ_2 we have either $A_{ij} = 0$ or $A_{ij} \ge 550 - 230\sqrt{5}$. The total sum is $\sum a_{ij} = 324$. $\square$

Corollary 9.7. *For all $p \in (0,1)^2$, the sequence $\{H^n(p)\}$ converges exponentially fast to the regular class.*

Proof: Let r be the regular class. Let $p_n = H^n(p)$ and let $f_n = f(p_n)$. Note that $\{f_n\}$ is non-decreasing. It follows directly from equation (9.8) that that the level sets $f^{-1}[f_1, 1]$ are compact. Hence $\{p_n\}$ stays within a compact subset of $(0,1)^2$. On a subsequence, we have $p_n \to q$ for some q in this compact set. By construction $f_n \to f(q)$. But if $q \neq r$ then we must have $f_n \to f(H(q)) > f(q)$, and this is a contradiction. Hence $q = r$. This completes the proof of convergence.

We already mentioned that r is an attracting fixed point for H, with multiplier $2\phi^{-5}$. Hence, the convergence happens exponentially fast. Once p_n is sufficiently close to r, we have $\|p_{n+1} - r\| < \|p_n - r\|/5$. Our bound comes from the fact that $2\phi^{-5} < 1/5$. □

9.5. The Action on the Boundary

For later purposes, it will be useful to know how H acts on the boundary of $[0,1]^2$. The reason we omit the point $(1,1)$ from our result is that H blows up at this point.

Lemma 9.8. *H maps $\partial[0,1]^2 - \{(1,1)\}$ into $(0,1)^2$.*

Proof: We will consider the two vertical sides. The case of the horizontal sides follows from symmetry. We compute

(9.14)

$$H(0,y) = \left(\frac{18}{30+y-y^2}, \frac{6-y}{2(5-y)}\right), \qquad H(1,y) = \left(\frac{(5-y)(3+y)}{(6-y)(4+y)}, \frac{6-y}{2(5-y)}\right)$$

One can see that both coordinates are in $(0,1)$ when $y \in [0,1]$. □

Remark: Note that the restriction of H to the vertical line $x = 1$ is completely defined, and maps $(1,1)$ into $(0,1)^2$.

9.6. Discussion

For $N \geq 6$. Let $f = O_n E_n$ be the product of the flag invariants. It is possible that Theorem 1.3 works on $\mathcal{C}_6$ without the claim about equality. We have already seen in §7.5 that the map H^*H has other fixed points in $\mathcal{C}_6$ besides the regular class. If $f(H(P)) > f(P)$ for all non-regular $P \in \mathcal{C}_6$ then by duality $f(H^*(P)) > f(P)$ for all non-regular $P \in \mathcal{C}_6$ But then $f(H^*H(P)) > f(P)$ for all non-regular $P \in \mathcal{C}_6$, and this contradicts the existence of these other fixed points.

On $\mathcal{C}_7$ there are some counterexamples to Theorem 1.3 and probably similar things go wrong for larger values of N. (I didn't test it.) In spite of this, it seems reasonable to make the following conjecture.

Conjecture 9.9. *For each N there is some even power $k = k_N$ such that*

$$f(H^k(P)) \geq f(P) \qquad \forall\ P \in \mathcal{C}_N \tag{9.15}$$

with equality if and only if P is the regular class. Probably one can even take $k = 2$ in all cases.

CHAPTER 10

The Basic Domains

10.1. The Space of Pentagons

For the rest of the monograph, we identify $\mathcal{P}$ with a subset of $\boldsymbol{R}^2$. The subset is $\boldsymbol{R}^2 - \Xi$, where Ξ is the union of the the following curves:

$$\{x = 0\}, \quad \{x = 1\}, \quad \{y = 0\}, \quad \{y = 1\}, \quad \{xy = 1\}. \tag{10.1}$$

$\boldsymbol{R}^2 - \Xi$ has 12 components, which we call $\mathcal{P}_j$ for $j = 1, \ldots, 10$ and $\mathcal{C}$ and $\mathcal{C}_*$. These last two stand for the convex classes and the star-convex classes respectively. Figure 10.1 shows the picture.

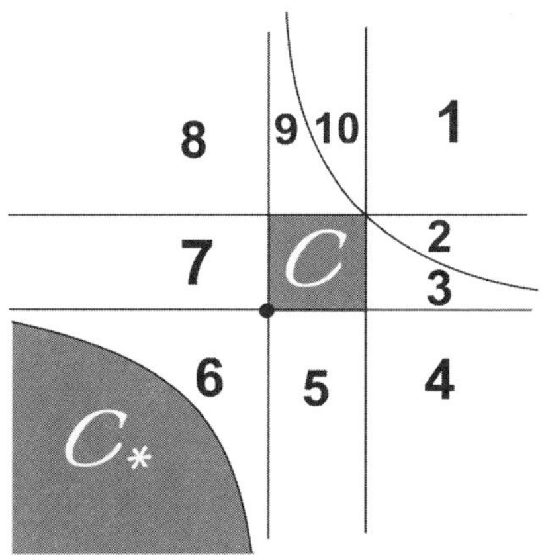

FIGURE 10.1. The components

The following result extends Lemma 9.1 and gives another proof of it.

LEMMA 10.1. *Ψ is a homeomorphism $\mathcal{P}$ onto $\boldsymbol{R}^2 - \Xi$.*

Proof: If $p \in \mathcal{P}$ then the points and lines of p are in general position, so that

$$\Psi(p) = (x, y) \in (\boldsymbol{R}^2 - \{0, 1\})^2.$$

The second coordinate of $G(x, y)$ is $(1-x)/(1-xy)$. So, if $xy = 1$ then $G(x, y) \notin \boldsymbol{R}^2$. Hence, $\Psi(p) \subset \boldsymbol{R}^2 - \Xi$.

By considering just 4 pentagons and their relabelings, it is a simple matter to see that the image of Ψ contains points in all 12 components of $\boldsymbol{R}^2 - \Xi$. Now Ψ is a local homeomorphism, and also if $\{p_n\}$ is a sequence of points exiting $\mathcal{P}$ then $\Psi(p_n)$ exits $\boldsymbol{R}^2 - \Xi$. Hence Ψ is a proper map from $\mathcal{P}$ into $\boldsymbol{R}^2 - \Xi$. But now it follows from Invariance of Domain that Ψ is a homeomorphism from $\mathcal{P}$ to $\boldsymbol{R}^2 - \Xi$. $\square$

10.2. The Action of the Gauss Group

The dihedral group acts as the group of dihedral relabelings of a labeled pentagons. With our identification, this action is given by the Gauss group Γ. Since we know in advance that Γ permutes the components of $\mathcal{P}$, we just have to check a few points to figure out the combinatorics of the action.

Γ preserves $\mathcal{C}$ and $\mathcal{C}_*$. The element $G \in \Gamma$, the Gauss recurrence from equation (3.16), has the following action on the other pieces: $1 \to 3 \to 5 \to 7 \to 9$ and $2 \to 10 \to 4 \to 6 \to 8$. The whole group Γ permutes the odd components and also permutes the even components.

The set of non-convex classes is denoted

$$\mathcal{N} = \mathcal{P} - \mathcal{C}. \tag{10.2}$$

Now we are going to define a fundamental domain for the action of Γ on $\mathcal{N}$. The reason for this weird domain will become clear in the next section. The domain $\mathcal{T}$ is a union of 3 unbounded regions in $\boldsymbol{R}^2$, but is is convenient to think of it as a connected subset of $(\boldsymbol{R} \cup \infty)^2$.

$\partial \mathcal{T}$ is the union of three curves:

- The unbounded segment of the line $y = x$ which joins $(1,1)$ to $(-\phi, -\phi)$. This segment is fixed by R. This is the curve labeled 0 in Figure 10.2.
- The unbounded subset of the line $x = 1$ which joins $(1,0)$ to $(1,1)$. This is the curve labeled 1 in Figure 10.2.

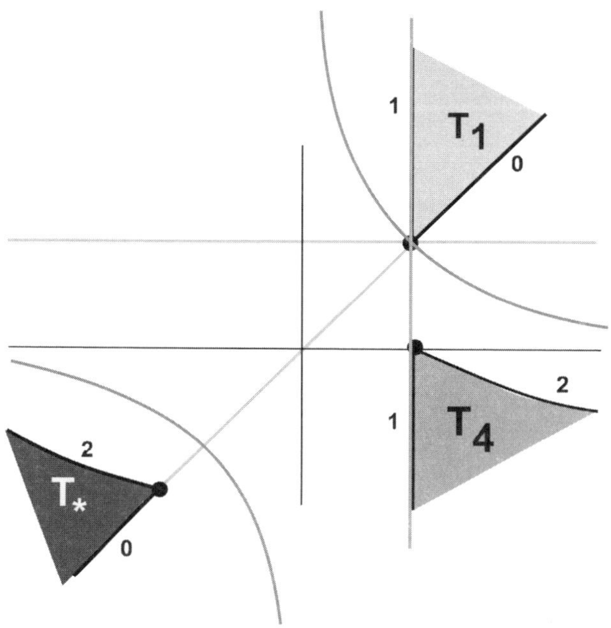

FIGURE 10.2. The fundamental domain $\mathcal{T}$ for Γ acting on $\mathcal{N}$

- The unbounded arc
$$\left(\frac{+\phi^{-1}+\phi^{+1}t}{-\phi^{-2}+\phi^{+2}t}, \frac{-\phi^{-0}+\phi^{+0}t}{+\phi^{-1}+\phi^{+1}t}\right), \qquad t \in [0,t].$$
This arc, which is labeled 2 in Figure 10.2, is fixed by $G^{-1}RG$. Here we have written $1 = \phi^{-0} = \phi^{+0}$ and generally tried to bring out the symmetry in the formula.

LEMMA 10.2. *$\mathcal{T}$ is a fundamental domain for the action of Γ on $\mathcal{N}$.*

Proof: We write
$$\mathcal{T} = T_1 \cup T_4 \cup T_*,$$
where $\mathcal{T}_k \subset \mathcal{P}_k$ for $k = 1, 4$ and $T_* \subset \mathcal{C}_*$. The element R preserves $\mathcal{P}_1$ and reflects across the diagonal. Hence T_1 is a fundamental domain for the action of R on $\mathcal{P}_1$. Since the boundary of T_4 contained in the interior of C_4 is fixed by $G^{-1}RG$, we see that T_4 is a fundamental domain for the action of $G^{-1}RG$ on $\mathcal{P}_4$. Finally, T_* is bounded by arcs which are fixed by R and $G^{-1}RG$. Hence T_* is a fundamental domain for the action of Γ on $\mathcal{C}_*$. Given how Γ permutes the components of $\mathcal{P}$, we see that T is a fundamental domain for the action on $\mathcal{N}$. □

10.3. Changing Coordinates

Here we repeat the formula for the map $\boldsymbol{B}$ in equation (1.2).

$$\boldsymbol{B}(x,y) = (b(x), b(y)), \qquad b(t) = \phi^3\left(\frac{\phi+t}{-1+\phi t}\right) \tag{10.3}$$

We define $\boldsymbol{T} = \boldsymbol{B}(\mathcal{T})$ and so on. $\mathcal{T}$ is a weird set, but we designed it to make $\boldsymbol{T}$ beautiful.

$\boldsymbol{T}$ is a Euclidean triangle with vertices

$$(0,0), \qquad (\phi^6, \phi^6), \qquad (\phi^6, -\phi^4). \tag{10.4}$$

Figure 10.3 shows the picture. The sides are labeled to indicate how $\boldsymbol{B}$ maps in $\partial\mathcal{T}$: The diagonal side of $\partial_0\boldsymbol{T}$. The vertical side is $\partial_1\boldsymbol{T}$. The other side is $\partial_2\boldsymbol{T}$. The triangle is shaded to indicate how $\boldsymbol{B}$ maps in the different pieces of $\mathcal{T}$. In $(\boldsymbol{R} \cup \infty)^2$ the regions T_1 and T_4 meet along $y = \infty$, which is mapped to $\boldsymbol{B}$ to $y = \phi^2$. Likewise, T_1 and T_* meet along $x = \infty$ which is mapped by $\boldsymbol{B}$ to $x = \phi^2$.

We set

$$\boldsymbol{G} = \boldsymbol{B}G\boldsymbol{B}^{-1}, \qquad \boldsymbol{R} = \boldsymbol{B}R\boldsymbol{B}^{-1}, \qquad \boldsymbol{\Gamma} = \langle \boldsymbol{G}, \boldsymbol{R}\rangle. \tag{10.5}$$

Note that $\boldsymbol{R}(x,y) = R(x,y) = (y,x)$.

Remark: We use boldface notation with the risk of causing some clash with conventionally used symbols. For instance, $\boldsymbol{N} = \boldsymbol{B}(\mathcal{N})$ usually stands for the set of natural numbers. We hope that this will not cause confusion. The context should be clear when we are considering objects associated to polygon spaces and maps and when we are considering numbers of various kinds.

Since $\mathcal{T}$ is a fundamental domain for the action of Γ on $\mathcal{N}$, we see that $\boldsymbol{T}$ is a fundamental domain for the action of $\boldsymbol{\Gamma}$ on $\boldsymbol{N}$. What we mean is that every $\boldsymbol{\Gamma}$ orbit $\boldsymbol{O}$ in $\boldsymbol{N}$ is such that $\boldsymbol{O} \cap \boldsymbol{T}$ is nonempty and contains at most one point in the interior of $\boldsymbol{T}$. Generically, $\boldsymbol{O} \cap \boldsymbol{T}$ is exactly one interior point of $\boldsymbol{T}$.

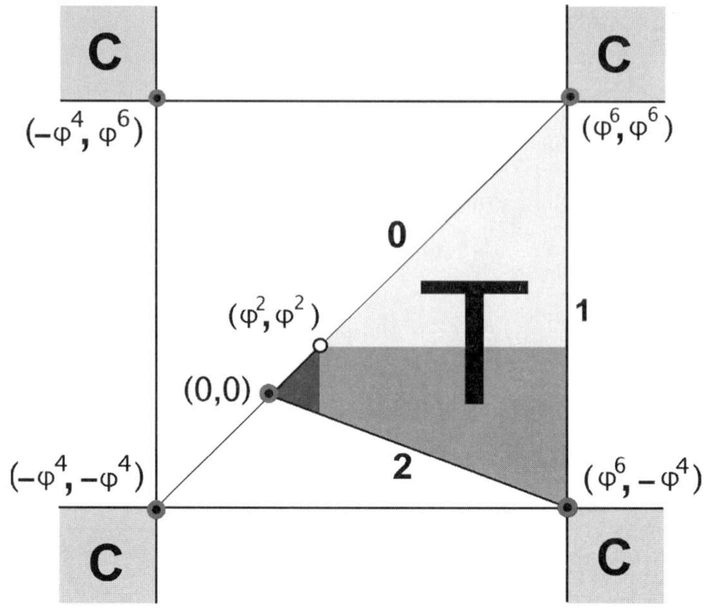

FIGURE 10.3. $\boldsymbol{T}$ and $\boldsymbol{C}$. Here $\boldsymbol{T}$ is the union of $\boldsymbol{B}(T_1)$ (light grey) and $\boldsymbol{B}(T_4)$ (medium grey) and $\boldsymbol{B}(T_*)$ (dark grey)

10.4. Convex and Star Convex Classes

The set $\boldsymbol{C}$, corresponding to the points of $\mathcal{C}$, is the union of the 4 unbounded sectors what we have suggested by the shading, together with the line at infinity. The map $\boldsymbol{B}$ blows up the regular class (ϕ^{-1}, ϕ^{-1}), so the domain $\boldsymbol{C}$ is a little bit funny. We will never deal directly with this domain.

To avoid ever dealing directly with $\boldsymbol{C}$, we introduce the map

$$\Theta = \boldsymbol{B}^{-1}\boldsymbol{H}. \tag{10.6}$$

The map Θ has the property that generically $\Theta(p) \in \mathcal{C}$ when $\boldsymbol{H}(p) \in \boldsymbol{C}$. We use the map Θ because $\mathcal{C}$ is a bounded domain. We always want to work with pairs (X, F) where X is a bounded domain in $\boldsymbol{R}^2$ and F is defined and finite on X.

The set $\mathbf{C}_*$, corresponds to the points of $\mathcal{P}$ which are star-convex. We have

$$\mathcal{C}_* \cap \boldsymbol{T} = \boldsymbol{T}_*. \tag{10.7}$$

Here $\boldsymbol{T}_*$, the dark-shaded triangle in Figure 10.3, has vertices

$$(0,0), \quad (\phi^2, \phi^2), \quad (\phi^2, -1). \tag{10.8}$$

Γ acts as a group of diffeomorphisms on $\mathcal{C}_*$ and the restriction of $\boldsymbol{B}$ to $\mathcal{C}_*$ is a diffeomorphism onto its image. Hence $\boldsymbol{\Gamma}$ acts as a group of diffeomorphisms on $\mathbf{C}_*$ and $\boldsymbol{T}_*$ is a fundamental domain for this action.

10.5. The Semigroup

We are interested in the semigroup generated by $\boldsymbol{H}$ and by the 10 elements of $\boldsymbol{\Gamma}$. Typically a semigroup of rational maps can be quite wild, but we first explain why this semigroup is practically the same thing as the semigroup of powers of $\boldsymbol{H}$.

LEMMA 10.3 (Commuting Maps). *$\boldsymbol{H}$ commutes with each element of $\boldsymbol{\Gamma}$,*

Proof: There are two proofs, one conceptual and one computational. Conceptually, all that is going on is that for $\gamma \in \boldsymbol{\Gamma}$, the composition $\gamma \boldsymbol{H} \gamma^{-1}$ is simply the 3-fold operation of changing coordinates, applying $\boldsymbol{H}$, and then changing back. The underlying geometric operation is the same as if we had just applied $\boldsymbol{H}$.

Computationally, it suffices to prove that $\boldsymbol{HR} = \boldsymbol{RH}$ and $\boldsymbol{HG} = \boldsymbol{GH}$, and this is checked by an explicit calculation that we omit. See §21.3 for the equations of all the maps involved. □

In light of the Commuting Maps Lemma, we can introduce a simple notation which records the name of every element in the semigroup. Define

$$\boldsymbol{H}_{ijk} = \boldsymbol{R}^i \boldsymbol{G}^j \boldsymbol{H}^k, \qquad i \in \{0,1\}, \qquad j \in \{0,1,2,3,4\}, \qquad k \in \{0,1\}. \tag{10.9}$$

The 10 maps $\boldsymbol{H}_{ij0}$ belong to $\boldsymbol{\Gamma}$ and the 10 maps $\boldsymbol{H}_{ij1}$ are what we call *primary maps*.

We let $\mathbf{S}$ denote the semigroup generated by the 10 maps of $\boldsymbol{\Gamma}$ and the 10 primary maps. Every element h of $\mathbf{S}$ which is not a generator can be written as

$$h = h_m \circ \cdots \circ h_1, \tag{10.10}$$

where the maps on the right are generators – some could be elements of $\boldsymbol{\Gamma}$ and some could be primary. In practice, our decompositions will have the property that the maps on the right are all primary.

Remarks:
(i) Our computer program lists out the formulas for the elements of $\boldsymbol{\Gamma}$ and also for the primary maps. That is, our computer program lists out the formulas for the 20 maps $\boldsymbol{H}_{ijk}$ with $k = 0, 1$.
(ii) All the maps in $\mathbf{S}$ have coefficients in $\boldsymbol{Z}[\sqrt{5}]$. However, we find it more convenient to work with the larger ring $R = \boldsymbol{Z}[1/2, \sqrt{5}, \sqrt{13}]$ because several of the important points associated to our maps have coordinates in R. See §8.3 and §21.4.
(iii) The reason we introduce $\mathbf{S}$ rather than just work with powers of $\boldsymbol{H}$ is twofold. First, $\mathcal{S}$ is a more canonical object. To define $\boldsymbol{H}$ we had to make a choice of coordinate system. $\mathbf{S}$ allows us to consider all coordinate systems at once. Second, we will sometimes switch elements of $\mathbf{S}$ in order to avoid maps which blow up. For instance, it might turn out that $\boldsymbol{H}$ blows up in some domain of interest to us while some other primary map does not, and so we will use the other primary map in place of $\boldsymbol{H}$. We will formalize this point of view presently.

We say that a generator h_k is *defined and finite* on some open set V if the denominators of the coordinate functions of h are nonzero throughout V. We say that h is *defined and finite* on an open set U if h_1 is defined and finite on U and h_2 is defined and finite on $h_1(U)$ and h_3 is defined and finite on $h_2 h_1(U)$, and so on. We say that h is *defined and finite* at a point if it is defined and finite in a neighborhood of that point. As we just remarked, we define things this way so as to avoid dealing with blowups.

10.6. A Global Point of View

We will take a more global point of view in Part 3 of the monograph, and here we discuss some of the underpinnings of that points of view.

Let M denote the space $(\boldsymbol{RP}^1)^2$ blown up at $(1,1)$, $(\infty,0)$ and $(0,\infty)$. Here we think of $\boldsymbol{RP}^1$ as $\boldsymbol{R}\cup\infty$. We will prove, in a step-by-step way, that Γ acts as a group of diffeomorphisms of M. The experienced algebraic geometer would doubtless find a more efficient proof of this fact. I work things out very concretely so that a reader unused to calculations with blowups can follow along.

We say that map $\Phi : X \to Y$ *acts nicely* if it is defined and continuous everywhere on X. Define the map

$$F = RG(x,y) = \left(\frac{1-x}{1-xy}, y\right) \tag{10.11}$$

LEMMA 10.4. *Γ acts as a group of diffeomorphisms on M provided that F acts nicely on M.*

Proof: Given the smooth nature of the blow-up construction and the action of Γ, we get the following result: If Γ acts as a group of homeomorphisms of M then Γ also acts as a group of diffeomorphisms. Note that Γ is generated R and F. Since we created M by blowing up an R-invariant manner, we know that R acts nicely on M. If F acts nicely on M then we have a set of generators which acts nicely on M. But then every element of G acts nicely. But then Γ acts as a group of homeomorphisms on M. $\square$

Our remaining goal is to show that $F : M \to M$ acts nicely. We will use symmetry to simplify the problem. Let $F^*(x,y) = (y/x, y)$. Let M^* denote the space obtained from $(\boldsymbol{RP}^1)^2$ by blowing up at $(0,0)$ and (∞,∞).

LEMMA 10.5. *F acts nicely on M provided that F^* acts nicely on M^*.*

Proof: Let $F_1 = F$ and $M_1 = M$. Let M_2 denote $(\boldsymbol{RP}^1)^2$ blown up at $(0,0)$, $(1,1)$, and (∞,∞). Let

$$F_2 = g_1 F_1 g_1^{-1}, \qquad g_1(x,y) = (1/x, y).$$

The map $g_1 : M_1 \to M_2$ is a homeomorphism. Hence, F_1 acts nicely on M_1 iff F_2 acts nicely on M_2.

Let M_3 denote $(\boldsymbol{RP}^1)^2$ blown up at $(-1,-1)$, $(0,0)$ and (∞,∞). Let

$$F_3 = g_2 F_2 g_2^{-1}, \qquad g_2(x,y) = (x-1, y-1).$$

The map $g_2 : M_2 \to M_3$ is a homeomorphism. F_2 acts nicely on M_2 iff F_3 acts nicely on M_3. We have

$$F_3(x,y) = \left(-\frac{y}{x}, y\right). \tag{10.12}$$

This map acts nicely at $(-1,-1)$ (without blowing up) so F_3 acts nicely on M_3 provided that F_3 acts nicely on M^*. But clearly F_3 acts nicely on M^* iff F^* acts nicely on M^*. $\square$

Now we simplify some more. Let $M^{**} \subset M^*$ denote the subset obtained by blowing up $\boldsymbol{R}^2$ at the origin.

LEMMA 10.6. *F^* acts nicely on M^* provided that $F^* : M^{**} \to M^*$ acts nicely.*

Proof: Consider the dihedral group D of homeomorphisms of M^* whose elements are the 4 maps $\mu_{\pm,\pm}(x,y) = (x^{\pm 1}, y^{\pm 1})$. Every orbit of D intersects M^{**}. Hence, it suffices to prove that the maps

$$\mu_{\pm,\pm}\ F^*\ \mu_{\pm,\pm} : M^{**} \to M$$

act nicely. These 4 maps are really just 2, namely $(x,y) \to (x/y, y)$ and $(x,y) \to (y/x, y)$. By symmetry it suffices to check the second of these on M^{**}, namely F^*. □

The next result finishes the proof.

LEMMA 10.7. *$F^* : M^{**} \to M^*$ acts nicely.*

Proof: F^* acts nicely on the subset of $\boldsymbol{R}^2$ where $x \neq 0$. When $x = 0$ and $y \neq 0$, we have $F^*(0,y) = (\infty, y)$, which is a perfectly good point in M^*. Finally, F^* is defined and continuous at the set $\widehat{O}$ of points of M^{**} lying over $(0,0)$. If $p_s \in \widehat{O}$ is the point representing the line of slope s through the origin, then $F^*(p_s) = (s, 0)$ when $s \neq 0$ and $F^*(p_0) = p_0$. So, $F^* : M^{**} \to M^*$ acts nicely. □

Now we explain what happens when we change coordinates. Let $\boldsymbol{M}$ be the space obtained by blowing up $(\boldsymbol{RP}^1)^2$ at the points

$$(10.13) \qquad (\phi^6, \phi^6) = \boldsymbol{B}(1,1), \qquad (\phi^2, -\phi^4) = \boldsymbol{B}(\infty, 0), \qquad (-\phi^4, \phi^2) = \boldsymbol{B}(0, \infty).$$

The map $\boldsymbol{B} : M \to \boldsymbol{M}$ is a diffeomorphism. Hence the group $\boldsymbol{\Gamma}$ acts as a group of diffeomorphisms of $\boldsymbol{M}$ and the space obtained from $\boldsymbol{T}$ by blowing up the vertex (ϕ^6, ϕ^6) is a fundamental domain for the action of $\boldsymbol{\Gamma}$ on $\boldsymbol{M} - \boldsymbol{C}$. Here $\boldsymbol{C} = \boldsymbol{B}(\mathcal{C})$ is the set of points representing convex pentagons. Figure 10.4 shows how the blowup turns the triangle $\boldsymbol{T}$ into a quadrilateral. Originally $\boldsymbol{T}$ and $\boldsymbol{C}$ touched at the point (ϕ^6, ϕ^6), but now they touch along a segment that is part of the blowup of this point. Moreover $\boldsymbol{C}$ changes from a piecewise analytic square to a piecewise analytic pentagon.

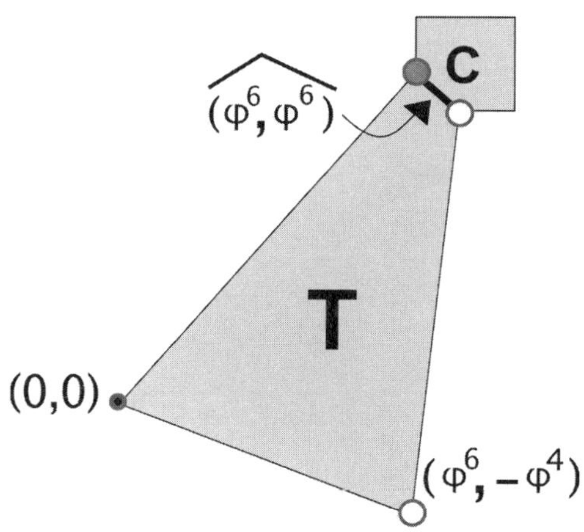

FIGURE 10.4. The Fundamental Domain blown up

Figure 10.5 shows the global picture. The region labeled (00) is (a homeomorphic copy of) $\boldsymbol{T}$. The region labeled (ij) is $\boldsymbol{H}_{ij0}(\boldsymbol{T})$. Both $\boldsymbol{H}_{100}$ and $\boldsymbol{H}_{120}$ act as (topological) reflections across the sides of $\boldsymbol{T}$ incident to the origin. The small pentagonal piece is (a homeomorphic copy of) $\boldsymbol{C}$, the set of points representing convex pentagons. It attaches to the big 15-gon along a 5 line segments, one of which is created when (ϕ^6, ϕ^6) is blown up. The big grey arrows indicate how the sides are identified.

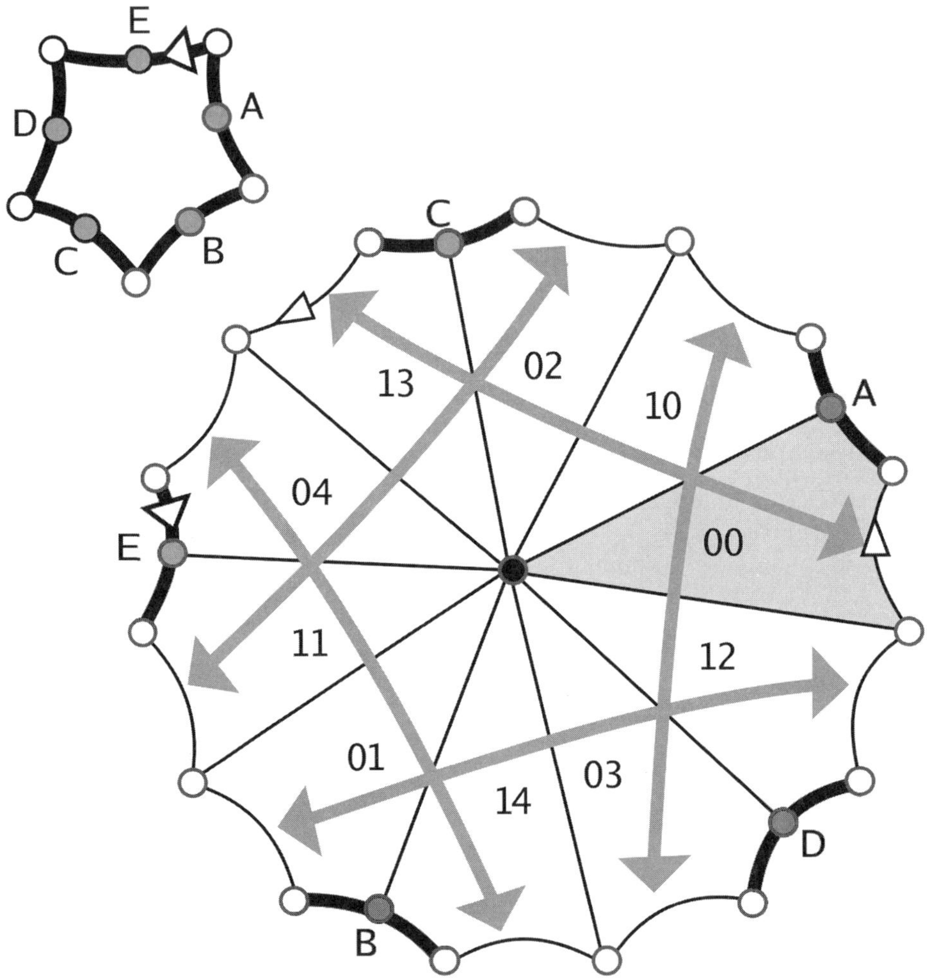

FIGURE 10.5. Global picture of the group action

CHAPTER 11

The Method of Positive Dominance

11.1. The Divide and Conquer Algorithm

Let S^o denote the interior of the set S.

In this chapter we explain the method of positive dominance. This is a computational method for showing that a polynomial is positive (or non-negative) on a polytope. I have not seen this method before in the literature. In the single variable case, the best method would probably be Sturm sequences. The advantage of the method of positive dominance is that it generalizes easily to higher dimensions.

Now we restrict ourselves to polynomials in $\boldsymbol{R}[x,y]$, though much of what we say generalizes to the k-variable case. The method of positive dominance is built on Lemma 9.4, which says the following about $F \in \boldsymbol{R}[x,y]$.

- If F is weak positive dominant, then $F > 0$ in $(0,1)^2$
- If F is positive dominant, then $F > 0$ in $[0,1]^2$

The notions of WPD and PD are based on various sums of the coefficients of F, as described in §9.3. We ignore the VWPD case of Lemma 9.4 because we don't need it. The converse of Lemma 9.4 is false. In this chapter we explain how to improve the criterion in Lemma 9.4 into a powerful machine.

We find it more convenient to work with triangles then squares, because it is easy to triangulate a polygon into triangles whereas one can rarely decompose a polygon into finitely many squares. Let Σ_0 be the triangle with vertices

$$(0,0), \quad (0,1), \quad (1,1). \tag{11.1}$$

Points $(x,y) \in \Sigma_0$ satisfy the inequalities $0 \le x \le y \le 1$. We call Σ_0 the *standard triangle*. There is a nice polynomial map $\Phi : [0,1]^2 \to \Sigma_0$ given by

$$\Phi(x,y) = (xy, y). \tag{11.2}$$

Φ is a homeomorphism between the interiors of the two sets, but collapses some points on the boundary of the square.

Now suppose that $P \in \boldsymbol{R}[x,y]$. Let Σ denote some triangle. We call the pair (P,Σ) *WPD* if there is some affine isomorphism $A : \Sigma_0 \to \Sigma$ such that the map

$$P \circ \Phi \circ A \tag{11.3}$$

is WPD. We make the same definition for PD. Note that there are 6 possible affine maps from Σ_0 to Σ, so there is some flexibility in our definition

LEMMA 11.1. *The following is true.*

(1) *If (P,Σ) is WPD then $P > 0$ on Σ^o.*
(2) *If (P,Σ) is PD then $P > 0$ on Σ.*

Proof: This follows straight from the definitions and Lemma 9.4. □

Remark: If the vertices of Σ have coordinates in a ring R, and P has coefficients in R, then $P \circ \Phi \circ A$ also has coefficients in R. This will be very important for the computational aspects of our computations.

Now we explain the *positive dominance algorithm*, or *PDA* for short. We would like to show that some polynomial P is positive on some polygon X. Before we start, we define our method for subdividing one triangle into 4 smaller ones. Figure 11.1 shows the construction. The extra vertices in the figure are at the midpoints of the edges of the large triangle.

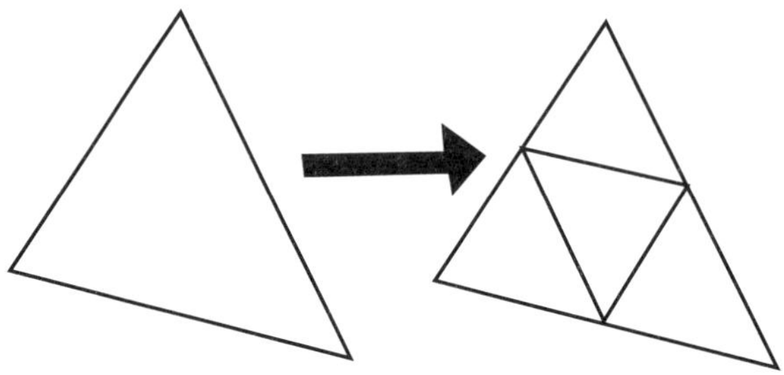

FIGURE 11.1. Subdivision of a triangle

Let $X_1, \ldots, X_m$ be the list of vertices of X. We always assume that X is star-shaped with respect to X_1, though this assumption is just for convenience. Here is the algorithm.

(1) We starts with a list LIST of triangles. The initial members of LIST are the triangles (X_1, X_i, X_{i+1}) for $i = 2, \ldots, m-1$.
(2) Let T be the first triangle on LIST. We delete T from LIST and check whether (P, T) is PD.
(3) Suppose (P, T) is PD. We return to Step 2 if LIST is nonempty and otherwise halt.
(4) Suppose (P, T) is not PD. We append to LIST the 4 triangles in the subdivision of T and then return to Step 2.

If the algorithm halts then it produces a partition of X into triangles on which P is PD. This implies that $P > 0$ on X.

The Weak Version: To get sharp results, we often need to use the algorithm above using WPD in place of PD. The reason is that sometimes the function of interest vanishes on some vertices of the domain. In this case the PDA will not work, and we need the weak version. We call this the WPDA. The WPDA has one annoying complication. The algorithm does not quite show that $P > 0$ on the given domain. It might *a priori* happen that $P = 0$ on some of the edges of the partition, even if these edges lie in the interior of the domain. In the few cases when we really need to get strict inequality with the WPDA, we will resort to one of two tricks which we describe below.

Edges Our tests are geared towards two dimensional polygons, but we can also use them to test whether a function is positive on an open or closed interval. We can verify that a function is positive on a closed interval $[p_1, p_2]$ by applying the

PDA to the degenerate triangle Σ whose vertices are p_1, p_2, p_2. All the definitions make sense. For positivity on an open interval (p_1, p_2) can use the WPDA, check that the partition has at least 3 intervals, and check that the function is PD on each interval of the partition lying in the interior of (p_1, p_2).

11.2. Positivity

So far, our discussion works in any setting where we can do exact arithmetic. Now we explain a feature of our calculations which is specific to the ring

$$R = \boldsymbol{Z}[1/2, \sqrt{5}, \sqrt{13}]. \tag{11.4}$$

Our algorithms are predicated on our ability to tell the sign of some element $a \in R$. In this section, we explain how to do this. I am indebted to Pat Hooper, who explained a very nice trick for doing this. Pat said that he, in turn, learned the trick from Vincent Delacroix.

As a first step, we replace a by some

$$b = 2^k a \in \boldsymbol{Z}[\sqrt{5}, \sqrt{13}]. \tag{11.5}$$

We take k as small as possible. As a warm-up, suppose first that $b \in \boldsymbol{Z}[\sqrt{5}]$. We write

$$b = m + n\sqrt{5}, \qquad m, n \in \boldsymbol{Z}. \tag{11.6}$$

The only nontrivial cases occur when m and n are both nonzero and have opposite signs. In these cases, we have the following rules.

- If $m > 0$ and $n < 0$ then b and $m^2 - 5n^2$ have the same sign.
- If $m < 0$ and $n > 0$ then b and $m^2 - 5n^2$ have opposite signs.

Both rules derive from the identity

$$b = \frac{m^2 - 5n^2}{m - n\sqrt{5}}.$$

Now we go back to the general case. We write

$$b = m + n\sqrt{13}, \qquad m, n \in \boldsymbol{Z}[\sqrt{5}]. \tag{11.7}$$

We already know how to tell the signs of m and n. The only nontrivial cases occur when both these numbers are nonzero and have opposite signs. In these cases, we have the following rules.

- If $m > 0$ and $n < 0$ then b and $m^2 - 13n^2$ have the same sign.
- If $m < 0$ and $n > 0$ then b and $m^2 - 13n^2$ have opposite signs.

The justification for these rules is as above. Since $m^2 - 13n^2 \in \boldsymbol{Z}[\sqrt{5}]$, we use the warm-up method to determine the sign of these numbers.

11.3. The Denominator Test

11.3.1. The Basic Test. Let X be a polygon and let F be a rational map.

$$F = (F_1, F_2), \qquad F_j = \frac{N_j}{D_j}, \tag{11.8}$$

We will use either the PDA to show that $D_1 > 0$ and $D_2 > 0$ on X or will use the WPDA (plus a trick) to show that $D > 0$ and $D_2 > 0$ on X minus some of its vertices. We call this test the *denominator test*.

11.3.2. The Subtraction Trick. We will encounter some situations where F blows up at one or two vertices of X. In all these cases $X^o \subset \boldsymbol{T}^o$, where $\boldsymbol{T}$ is our fundamental domain, and the offending vertices will lie in the two edges $\partial_0 \boldsymbol{T}$ and $\partial_1 \boldsymbol{T}$ of $\boldsymbol{T}$ which are incident to the origin. See Figure 10.3. We define

$$g(x,y) = \phi^{-9}(x-y)(\phi^2 x + y), \tag{11.9}$$

This function is positive on $\boldsymbol{T} - \partial_0 \boldsymbol{T} - \partial_2 \boldsymbol{T}$. We take the following 3 steps.

(1) We use the WPDA to check that $D_j - g \geq 0$ on X^o and this shows that $D_j > 0$ on $X - \partial_0 \boldsymbol{T} - \partial_2 \boldsymbol{T}$.
(2) We check that D_j is WPD on each edge of X that lies in $\partial_0 \boldsymbol{T} \cup \partial_2 \boldsymbol{T}$ This shows that $D_j > 0$ on the interiors of such edges.
(3) We check that $D_j > 0$ on any vertices of X which lie in $\partial_0 \boldsymbol{T} \cup \partial_2 \boldsymbol{T}$ and are not the offending vertices.

For later reference, we call this the *subtraction trick*.

11.3.3. The Variation Trick. There will be two cases where the subtraction trick does not work. In these cases, we produce two new domains Y and Z by varying a vertex p of X in such a way that $X \subset Y$ and $X \subset Z$. For each j we use the WPDA (three times) to show that $D_j > 0$ on a partition X^* (respectively Y^*, Z^*) relative to X (respectively Y, Z). The partitions are combinatorially identical. We perturb p so that there is no point of X^o which lies in an edge of all three partitions.

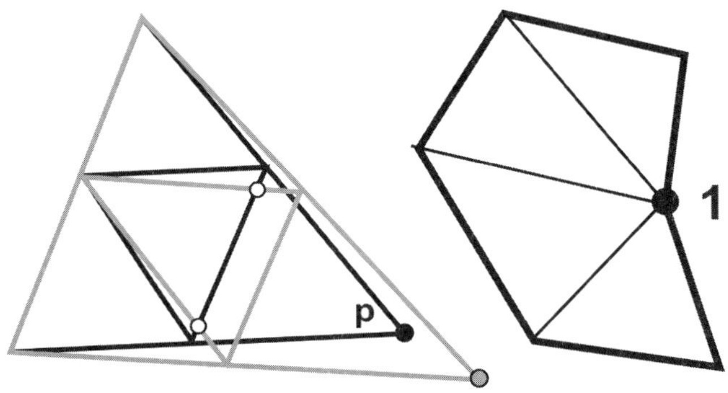

FIGURE 11.2. Constructions for the variation trick

The left side of Figure 11.2 shows how the partitions X^* and Y^* intersect when X is a triangle. The two highlighted points are the only points in X^o common to the edges of X^* and Y^*. In general, there would be finitely many such points. When we also consider Z^*, there are no points in common.

Our argument shows that $D_j > 0$ on X^o. We then deal with the edges and vertices as in the subtraction trick. For later reference, we call this the *variation trick*.

There is more we want to say about the variation trick. When our domain X is a polygon with more than 3 sides, we triangulate X by connecting all the vertices to the vertex x_1 of X. When we do the variation trick, we vary x_1, so that each triangular domain is affected. We do enough variations so that we arrange the situation above (namely $X \subset Y, Z$) for each of the triangles in the triangulation.

11.4. The Area Test

Let X' either be X or X minus some of its vertices. Suppose we already know that F is defined and finite on X', and we want to show that $\pm\det(dF) \geq C$ on X' for some sign choice.

Remark: We want to make it clear that we are not testing the two inequalities $\det(dF) \geq C$ and $-\det(dF) \geq C$. This is almost absurd. Rather, $\pm\det(dF)$ denotes the choice of $+\det(dF)$ or $-\det(dF)$ that we have decided to test.

We write

$$\pm\det(dF) - C = \frac{\Delta}{D_1^1 D_2^2}, \tag{11.10}$$

where Δ is a polynomial that depends on F and C. We then use the WPDA to check that $\Delta > 0$ on X^o. If follows from continuity that $\Delta \geq 0$ on X'. Hence $\pm\det(dF) \geq C$ on X'. We call all this the *area test.*

Our main use of the area test will be to show that $\pm\det(dF) > 0$ on X'. This will guarantee that F is a local diffeomorphism on X', because dF never vanishes. In all but 4 cases we will use the PDA to show that $\pm\det(dF) > 0$ on X. In 3 of the 4 exceptional cases, we will use the WPDA to show that $\pm\det(dF) - \phi^{-2} \geq 0$ on X^o, and this shows that $\pm\det(dF) > 0$ on X'. In the remaining exceptional case, we will use the WPDA and the variation trick to show that $\det(dF) > 0$ on X'.

11.5. The Expansion Test

Remark: We only use the PDA for the tests in this section, so there are no tricks we need to use.

Let X and F be as in the previous two sections. We explain how to check that $F : X \to \boldsymbol{R}^2$ is an expanding map, and hence a diffeomorphism onto its image. The basic idea is to compare the L_2 norm of the differential with the determinant of the differential.

We find it convenient to use the variables (x_1, x_2) for points in $\boldsymbol{R}^2$. Given two positive constants C_1 and C_2, we introduce the polynomials Ψ_1 and Ψ_2, defined as follows:

$$\frac{\Psi_j}{D_j^2} = C_j - \left(\frac{\partial F_j}{\partial x_1}\right)^2 - \left(\frac{\partial F_j}{\partial x_2}\right)^2. \tag{11.11}$$

We use the PDA to check that $\Psi_j > 0$ on X. This proves that

$$\sum_{i=1}^{2}\sum_{j=1}^{2}\left(\frac{\partial F_i}{\partial x_j}\right)^2 < C_1 + C_2. \tag{11.12}$$

on X.

We recognize the left hand side as the square of the L_2 norm of dF. Thus we can say that

$$\|dF\|_2 < \sqrt{C_1 + C_2}. \tag{11.13}$$

Geometrically, this means that dF expands no vector more than $\sqrt{C_1+C_2}$ on X.

Remark: It would be simpler mathematically to directly bound $\|dF\|_2$ without splitting things up into two polynomials, but computationally it is much worse. The sum of the squares of dF is a truly enormous rational function. The key advantage to our approach is that the two functions $(\partial_1 F_j)^2$ and $(\partial_2 F_j)^2$ have the same denominator, so when their squares are added, the size of the polynomial is not much larger than the terms separately. This fact about the denominators comes from the quotient rule

$$\partial\left(\frac{P}{Q}\right) = \frac{\partial P\, Q - \partial Q\, P}{Q^2}.$$

There are no derivatives in the denominator, so it doesn't matter which partial derivative we use.

LEMMA 11.2. *Suppose, on X, that $|\det(dF)| > C_0$ and $\|dF\|_2 < \sqrt{C_1+C_2}$. If $\sqrt{C_1+C_2} < C_0$, then F is an expanding map on X.*

Proof: Let V be any unit tangent vector, based at some point where the inequalities hold. Let W be a unit vector which is orthogonal to V. Let Π be the unit square whose sides are V and W. The parallelogram $dF(\Pi)$ has area at least C_0. On the other hand, $\|dF(W)\| \leq \sqrt{C_1+C_2} < C_0$. But the area of $dF(\Pi)$ is at least $\|dF(V)\|\|dF(W)\|$. This forces $\|dF(V)\| > 1$, as desired. $\square$

Under these circumstances, we call the quintuple (X, F, C_0, C_1, C_2) *expansive*. If such a quintuple is expansive, then F is a diffeomorphism on X and an expanding map.

11.6. The Confinement Test

Let F and X' be as in previous sections. Suppose we have a quadruple (F, X', p, q), where F is defined and finite on X' and $p, q \in \boldsymbol{R}^2$ are distinct points. Let $L = (pq)$ be the line containing p and q. Here we describe how we verify that $F(X')$ lies in one of the two closed half-planes bounded by L.

Let $I(x, y) = (-y, x)$ denote 90 degree rotation. Define $\Lambda(x, y) = Ax + By + C$, where

$$r = p + I(q-p), \qquad A = r_1 - p_1, \qquad B = r_2 - p_2, \qquad C = p \cdot (p - r). \tag{11.14}$$

Here $p = (p_1, p_2)$, etc. A short calculation, which we omit, shows that $\Lambda = 0$ on L.

We consider the rational function

$$\Lambda \circ F = \frac{Z}{D_1 D_2}, \qquad Z = Z_{F,X,p,q} \tag{11.15}$$

We use the WPDA to check that $D_1, D_2 \geq 0$ on X' and either $Z \geq 0$ on X' or $Z \leq 0$ on X'. In either case, this forces $F(X')$ to lie in one of the two closed half-planes bounded by L. The point here is that $F(X')$ crossed L, then Z would take on both signs on X'. We call this this the *half-plane test*.

The half-plane test is the basis for one of our two main geometric tests. Let (F, X', Y) be as above, with Y a convex polygon. Here we describe a test which verifies that

$$F(X') \subset Y \tag{11.16}$$

when F is defined and finite on X'. Let $v_1, \ldots, v_k$ be the successive vertices of Y. We apply the half-plane test to the quadruples (F, X, v_i, v_{i+1}) for each $i = 1, \ldots, k$ (with indices taken cyclically) and also we check that F maps some point of X into the interior of Y.

The Strong Variant: There will be two special situations where we need to know that $F(X) \subset Y^o$. In this case we will use the PDA on X for all the tests mentioned in this section.

11.7. The Exclusion Test

Remark: For the test in this section we only use the WPDA. We do not care about strict inequalities here.

Here is the second of our big geometric tests. Suppose that (F, X', Y) is a triple where X and Y are polygons and Y is convex. Suppose that F is defined and finite on X'. Here we explain how to prove that

$$F(X') \cap Y^o = \emptyset. \tag{11.17}$$

In other words, we are excluding $F(X)$ from Y^o. We call this the *exclusion test*.

Let (F, T, p, q) be a quadruple, where T is a triangle $p, q \in \boldsymbol{R}^2$ are two distinct points. We call (F, T, p, q) *nice* if one of the two polynomials $\pm Z_{F,T,p,q}$ from equation (11.15) is defined and non-negative on a dense subset of T, and there is an interior point $t \in T$ such that $F(t)$ lies on the other side of $L = (pq)$ as Y. Under these circumstances, L separates $F(T)$ from Y^o.

Consider the following algorithm:

(1) Start with a list LIST of triangles. Initially, LIST contains the triangles coming from the subdivision of X.
(2) Let T be the first triangle on LIST. We delete T from LIST and then test whether (P, T, Y_i, Y_{i+1}) is nice for some i.
(3) Suppose some quadruple is nice. We go back to Step 2 if L is nonempty and otherwise halt.
(4) Suppose no quadruple is nice. We append to L the 4 triangles obtained from subdividing T, then go back to Step 2.

If this algorithm halts, it produces a partition of X into triangles $\{T_k\}$ with the property that each image $F(T_k)$ is separated from Y^o by a line extending some edge of Y. This implies that $F(X') \cap Y^o = \emptyset$.

11.8. The Cone Test

We say that a *cone* C in $\boldsymbol{R}^2$ is the region in $\boldsymbol{R}^2$ bounded by two lines through the origin of unequal slope. If the two lines are perpendicular, we call the cone *right angled*. For instance, the cone $\vee$ from the previous section is a right-angled cone. If C strictly contains a right-angled cone, we call C *obtuse*. If C is strictly contained in a right-angled cone, we call C *acute*. We define $C^\perp$ to be the cone bounded by the lines perpendicular to the lines bounding C. We have $C = C^\perp$ if C is a right cone. If C is acute (respectively obtuse) then $C^\perp$ is obtuse (respectively acute).

Given a cone C, we say that a *basis* for C is a pair nonzero vectors V_1 and V_2 contained in the lines bounding C which have the following property: A vector W lies in C if and only if W is either a non-negative or a non-positive linear combination of V_1 and V_2. For instance, if $\vee$ is the cone of vectors having slope whose absolute value is at least 1, then then $\{(1,1),\ (-1,1)\}$ is a basis for $\vee$. Here $\vee$ is a right-angled cone.

Let X' be as above. Suppose that $f : X' \to \boldsymbol{R}^2$ is smooth and well defined on X'. Suppose that C and C' are cones. Here is how we check that $df_p(C) \subset C'$ for all $p \in X'$. We first choose a basis $\{V_1, V_2\}$ for C and a basis $\{V_1', V_2'\}$ for $(C')^\perp$.

In all cases, we use the WPDA to check that

$$df_p(V_i) \cdot V_j' \geq \epsilon, \qquad \forall i, j, \qquad \forall p \in X'. \tag{11.18}$$

for some $\epsilon > 0$. In practice we take $\epsilon = \phi^{-2}$. By choosing a positive constant, we get the stronger result that df_p maps C strictly inside C' – i.e., into a strictly smaller cone. Moreover, $df_p(V)$ is nonzero for all nonzero $V \in C$. Almost all our calculations use the right-angled case.

As a special case, we might want to show that $df_p(V) \subset C'$ for some particular vector V. In this case we set $V_1 = V_2 = V$ and make the above check. We also call this the cone test.

Cone Notation: Here is our notation for cones. We let $\vee(a,b)$ denote the closed cone of lines through the origin whose slope increases from a to b, possibly passing through ∞. For instance $\vee(1,-1)$ is the standard light cone consisting of lines through the origin whose slope is at least 1 in absolute value. $\vee(-\infty,-1/2)$ is the cone of lines through the origin whose slope is less or equal to $-1/2$. The only time we will deviate from this notation is that we will sometimes write $\vee = \vee(1,-1)$ for the standard light cone.

11.9. The Stretch Test

We will have one occasion to check that the differential df stretches distances by a factor of 5 for all vectors $V \in \vee$, the standard cone. Since we only use this test once, we will be very specific about it. However, the reader could imagine variants. The following lemma justifies our test.

LEMMA 11.3. *Let $T : \boldsymbol{R}^2 \to \boldsymbol{R}^2$ be linear, and let $\{v_1, v_2\}$ be an orthonormal basis. If $\|T(v_j)\| \geq \lambda$ for $j = 1,2$ and $T(v_1)$ makes an acute angle with $T(v_2)$ then $\|T(w)\| \geq \lambda \|w\|$ for any w which is a positive combination of v_1 and v_2.*

Proof: By scaling, it suffices to prove the case when $\lambda = 1$. Write $w = a_1 v_1 + a_2 b_2$. Let $v_j' = T(v_j)$ and $w' = T(w)$. We have

$$\|w'\| = a_1^2 \|v_1'\|^2 + a_2^2 \|v_2'\|^2 + 2a_1 a_2 (v_1' \cdot v_2') \geq a_1^2 + a_2^2 = \|w\|^2.$$

This does it. $\square$

For the stretch test, we set $V_1 = (1,1)$ and $V_2 = (-1,1)$ and for each $j = 1,2$ we use the PDA to check one of two statements.

- Setting $(X,Y) = df(V_j)$ we have $Y - 8 > 0$.
- Setting $(X,Y) = df(V_j)$ we have $-Y - 8 > 0$.

In either case we see that $\|df(V)\|/\|V\| \geq 8/\sqrt{2} > 5$. So, Lemma 11.3 applies to the orthonormal basis $\{V_1/\sqrt{2}, V_2/\sqrt{2}\}$. We call this the *stretch test*.

CHAPTER 12

The Cantor Set

12.1. Overview

The goal of chapter is to prove Theorem 1.5. In other words, we will construct the invariant Cantor set $\mathcal{J}C$ and analyze its properties. We will work in the $\boldsymbol{B}$-coordinates until the very end. The $\boldsymbol{B}$ coordinates are defined in equation (1.2) and discussed in depth in §10.

Recall from §10 that $\boldsymbol{T}$ is our triangular fundamental domain for the Gauss group $\boldsymbol{\Gamma}$. We will find a topological disk $\boldsymbol{D}$ and a union $\boldsymbol{D}_1 \cup \cdots \cup \boldsymbol{D}_6$ of 6 pairwise disjoint topological disks with the following property. For each j, the map $\boldsymbol{H} : \boldsymbol{D}_j \to \boldsymbol{R}^2$ is a diffeomorphism onto its image, and $\boldsymbol{D} \subset \boldsymbol{H}(\boldsymbol{D}_j)$. We will use this property to create the Cantor set, as discussed in §6.2. When we analyze the metrical and measure-theoretic properties of our construction we will get all the claims in Theorem 1.5.

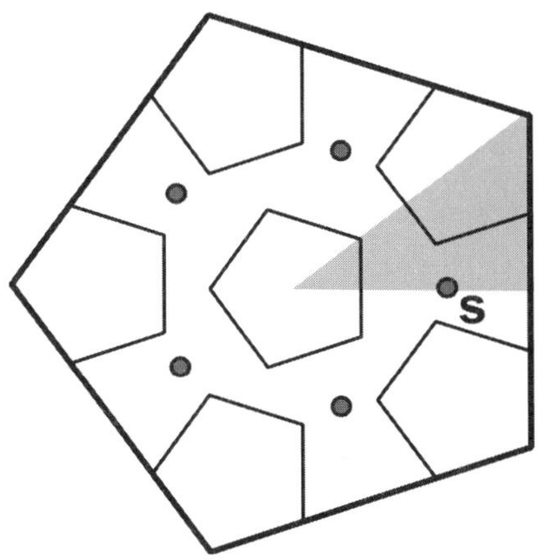

FIGURE 12.1. Schematic picture of the disks

Figure 12.1 shows a schematic picture of the disks. It turns out that there is a special orbit $\widehat{\Gamma}(\mathbf{s})$ on which the primary maps in the semigroup $\mathbf{S}$ blow up. This orbit is also shown schematically in Figure 12.1. It plays a special role in our constructions.

When we refer to pieces in our partition we will always use the symbols $\clubsuit Q$, $\clubsuit R$, etc. When we refer to polygonal subsets of the moduli space which interact with the partition but are not part of the partition, we use the notation $\spadesuit Y$, $\spadesuit Z$, etc. Finally, when we refer to any subset of the Julia set in $\boldsymbol{B}$-coordinates, we use the notation $\heartsuit A$, $\heartsuit B$, etc. These conventions are designed to avoid conflicts in notation, and to jog the reader's memory about the meaning of the various symbols we use. In §21 we list the vertices of our partition pieces.

12.2. The Big Disk

The main partition piece in this chapter is $\clubsuit Q$, a convex quadrilateral obtained from the triangle with vertices

$$(0,0), \qquad (1,1), \qquad (\phi, -1/\phi).$$

by chopping off (for technical reasons) a tiny corner near the vertex $(\phi, -1/\phi)$. The fundamental domain $\boldsymbol{T}$ shares two sides with $\clubsuit Q$. Our big disk $\boldsymbol{D}$ is

$$\boldsymbol{D} = \bigcup_{\gamma \in \boldsymbol{\Gamma}} \gamma(\clubsuit Q). \tag{12.1}$$

Figure 12.2 shows how $\clubsuit Q$ sits inside $\boldsymbol{T}$, our fundamental domain. The lightly shaded triangle is the set of star convex points in $\boldsymbol{T}$. Figure 12.3 below shows $\boldsymbol{D}$.

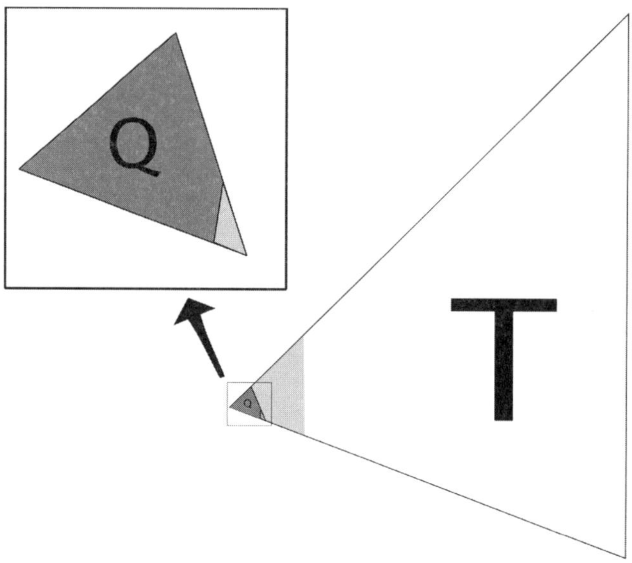

FIGURE 12.2. The sets $\clubsuit Q$ and $\boldsymbol{T}$

LEMMA 12.1. *$\boldsymbol{D}$ is a topological disk with piecewise analytic boundary. Moreover, $\boldsymbol{\Gamma}$ acts as a group of diffeomorphisms on $\boldsymbol{D}$.*

Proof: Since $\clubsuit Q$ lies entirely inside the set of points representing star convex pentagons, the whole disk $\boldsymbol{D}$ has this property as well. Hence $\boldsymbol{\Gamma}$ acts as a group of analytic diffeomorphisms of $\boldsymbol{D}$. Now we check that $\boldsymbol{D}$ is a topological disk. First, the sides of $\clubsuit Q$ incident to the origin are each fixed by an element of $\boldsymbol{\Gamma}$. Second, using the weak exclusion test, we check that $\gamma(\clubsuit Q)$ and $\clubsuit Q$ have disjoint interiors for all nontrivial $\gamma \in \boldsymbol{\Gamma}$. These two facts show that the $\boldsymbol{\Gamma}$ images of $\clubsuit Q$ fit together like slices in a pizza to make a topological disk. The boundary is clearly piecewise analytic. □

Figure 12.3 shows an accurate picture of $\boldsymbol{D}$.

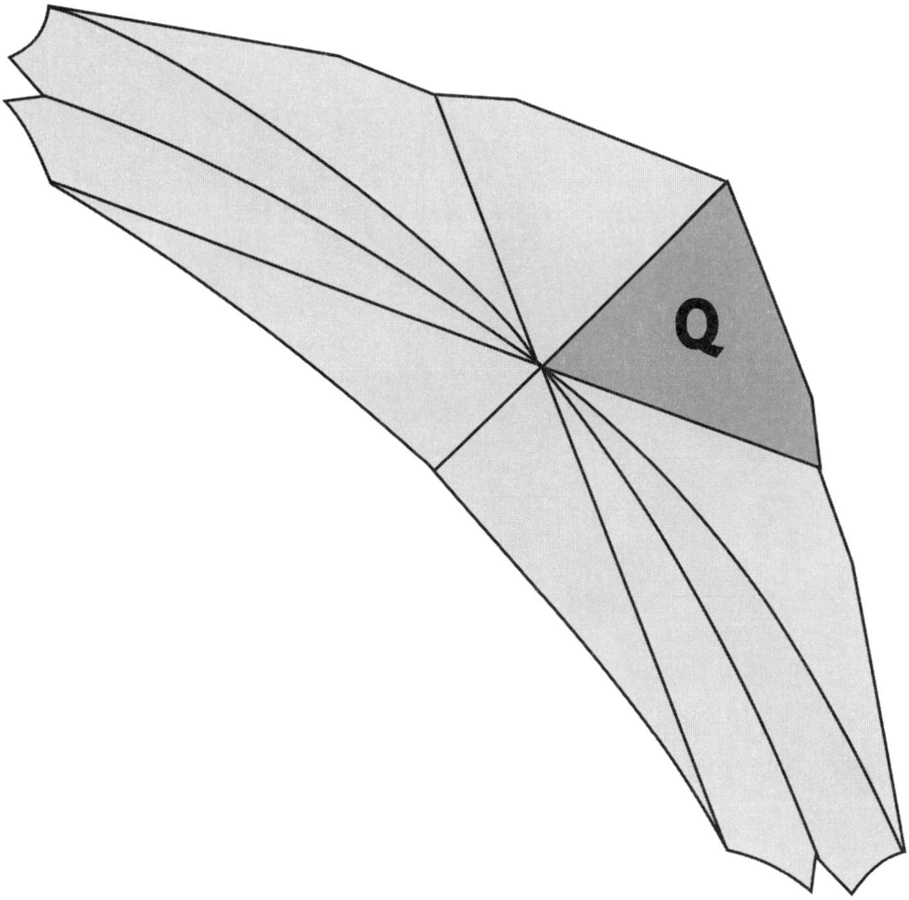

FIGURE 12.3. The disk $\boldsymbol{D}$

12.3. The Six Small Disks

In order to define our 6 smaller disks, we introduce partitions

$$\clubsuit Q = \clubsuit R \cup \clubsuit S \cup \clubsuit T, \tag{12.2}$$

$$\clubsuit S = \clubsuit S_1 \cup \clubsuit S_2 \cup \clubsuit S_3. \tag{12.3}$$

$$\clubsuit T = \clubsuit T_1 \cup \clubsuit T_2 \cup \clubsuit T_3 \cup \clubsuit T_4 \cup \clubsuit T_5. \tag{12.4}$$

We explain in §12.8 why our partition has the structure it does. Basically, the pieces $\clubsuit R$, $\clubsuit S$, and $\clubsuit T$ function as single units, but in order to make some technical estimates we split these pieces up.

We list all the vertices of our partition in §21.5. The point $\mathbf{s}$ turns out to be the only point in $\clubsuit Q$ where all the primary maps blow up. The point $\mathbf{t}$ turns out to be the only point in the closure of the Julia set on the diagonal edge of $\clubsuit S_1$. (We ignore $\mathbf{t}$ in this chapter.) It turns out that

$$\bigcup_{\gamma \in \boldsymbol{\Gamma}} \gamma(\clubsuit R \cup \clubsuit T) \tag{12.5}$$

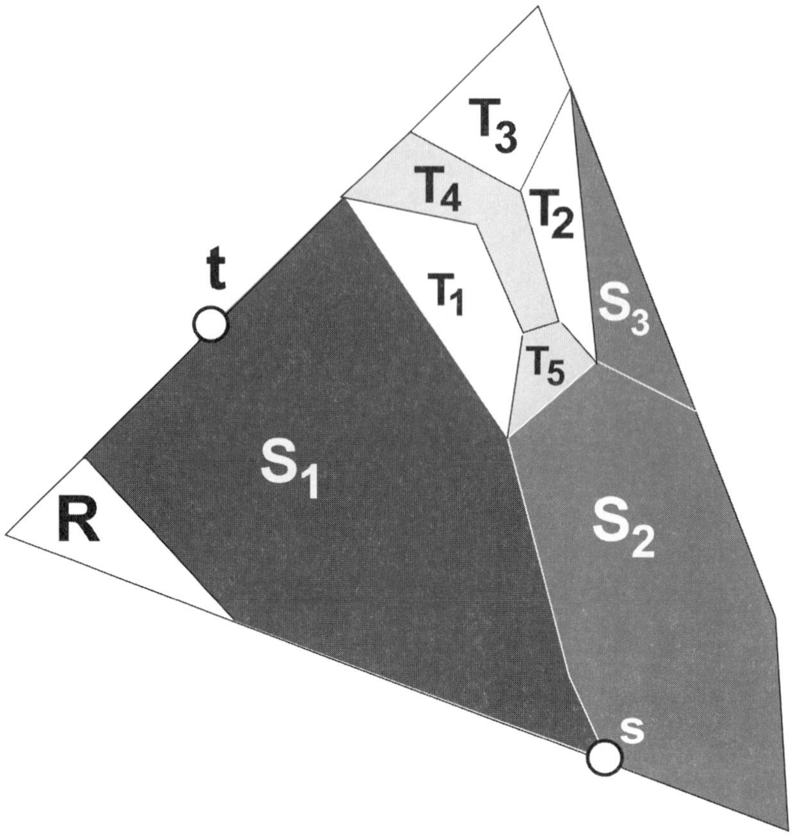

FIGURE 12.4. The partition of $\clubsuit Q$

is the union of 6 topological disks. These disks are:

$$\boldsymbol{D}_1 = \mathbf{\Gamma}(\clubsuit R), \qquad \boldsymbol{D}_k = \mathbf{G}^{k-2}(\clubsuit T \cup \boldsymbol{R}(\clubsuit T)), \quad k = 2,3,4,5,6. \tag{12.6}$$

Here $\boldsymbol{R}$ is reflection in the main diagonal and $\boldsymbol{G} = \boldsymbol{B}G\boldsymbol{B}^{-1}$ where G is the Gauss Recurrence from equation (3.16).

LEMMA 12.2. *$\boldsymbol{D}_1, \ldots, \boldsymbol{D}_6$ are pairwise disjoint topological disks.*

Proof: By symmetry it suffices to prove that $\boldsymbol{D}_1$ and $\boldsymbol{D}_2$ are topological disks, that $\boldsymbol{D}_1$ and $\boldsymbol{D}_2$ are disjoint, and that $\boldsymbol{D}_2$ and $\boldsymbol{D}_j$ are disjoint for all $j \in \{3,4,5,6\}$.

The fact that $\boldsymbol{D}_1$ is a topological disk has the same proof as the result for $\boldsymbol{D}$. The result for $\boldsymbol{D}_2$ is fairly obvious: $\clubsuit T$ is a polygon lying beneath the diagonal, except for one edge that is in the diagonal, and R reflects $\clubsuit T$ in the diagonal.

The orbit $\mathbf{\Gamma}(\clubsuit S)$ separates $\boldsymbol{D}_1$ from $\boldsymbol{D}_2$. For $j \in \{3,4,5,6\}$, the disks $\boldsymbol{D}_2$ and $\boldsymbol{D}_j$ are disjoint because $\boldsymbol{D}_2$ is entirely contained in $\clubsuit Q \cup R(\clubsuit Q)$, a union of two consecutive fundamental domains for $\mathbf{\Gamma}$ acting on $\boldsymbol{D}$. $\square$

12.4. The Diffeomorphism Property

Now that we have our disks, we prove that the restriction of any primary element to any disk $\boldsymbol{D}_k$ is a diffeomorphism onto its image which maps $\boldsymbol{D}_k$ over $\boldsymbol{D}$. Call this the *diffeomorphism property*. We also prove that the primary maps, where defined and finite, carry points of $\boldsymbol{D} - \bigcup \boldsymbol{D}_k$ outside of $\boldsymbol{D}$.

Note the following symmetry:

- Any 2 primary maps agree up to post-composition by an element of $\boldsymbol{\Gamma}$.
- $\boldsymbol{D}$ is invariant under $\boldsymbol{\Gamma}$.
- $\boldsymbol{\Gamma}$ acts as a group of diffeomorphisms on the space $\boldsymbol{C}_*$ of star convex classes, as discussed in §10.4.

Thanks to this symmetry, to establish the diffeomorphism property for all $\boldsymbol{D}_k$ it suffices to prove for $k = 1, 2$ that there is a primary map h_k such that

- $h_k : \boldsymbol{D}_k \to \boldsymbol{R}^2$ is a diffeomorphism onto the image.
- $\boldsymbol{D} \subset h_k(\boldsymbol{D}_k)$.
- $h_k(\boldsymbol{D}_k) \subset \boldsymbol{C}_*$.

If h_k has all these properties, we call h_k *good* with respect to $\boldsymbol{D}_k$.

LEMMA 12.3. *$\boldsymbol{H}_{111}$ is good with respect to $\boldsymbol{D}_1$.*

Proof: Let $h = \boldsymbol{H}_{111}$. Let e be the edge of $\clubsuit R$ that is not an edge of $\clubsuit Q$.

(1) We use the denominator test to check that h is defined and finite on $\clubsuit R$.
(2) We check that $(h, \clubsuit R, 26, 120, 40)$ is expansive. Hence h is an expanding diffeomorphism on $\clubsuit R$.
(3) We use the exclusion test to check that $f(\clubsuit R) \cap (\boldsymbol{T} - \boldsymbol{C}_*) = \emptyset$ for all 10 primary elements f. Hence $h(\boldsymbol{D}_1) \subset \boldsymbol{C}_*$.
(4) We use the exclusion test to show that $h(e) \cap \clubsuit Q = \emptyset$. Hence h maps e outside of $\clubsuit Q$.

Figure 12.5 shows an accurate picture.

The sets $\clubsuit R$ and $\clubsuit Q$ both have the vertex $(0, 0)$. The sides of $\clubsuit R$ and $\clubsuit Q$ incident to $(0, 0)$ lie in the sides of $\boldsymbol{T}$ incident to $(0, 0)$. Moreover, h maps the sides of $\clubsuit R$ into the sides of $\boldsymbol{T}$ (switching top and bottom). Our result now follows from dihedral symmetry and from all the structure we have established. □

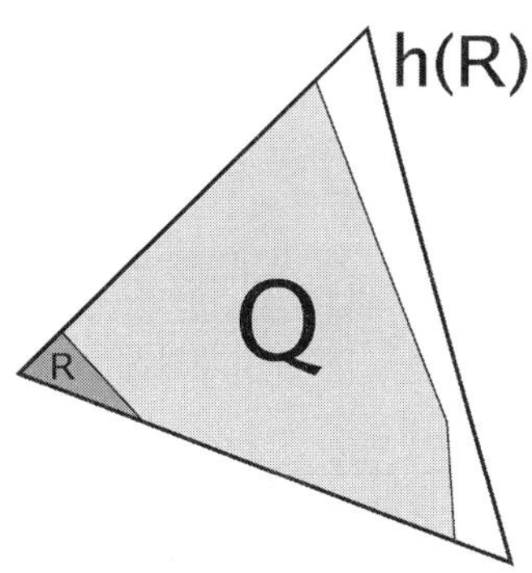

FIGURE 12.5. $\clubsuit R$ and $\clubsuit Q$ and $h(\clubsuit R)$

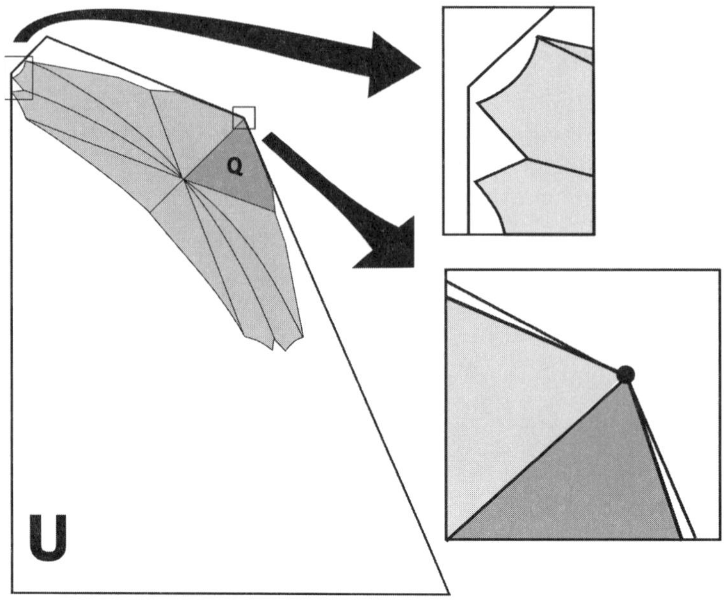

FIGURE 12.6. $\boldsymbol{D}$ and $\spadesuit U$

For our next two results, we introduce the convex pentagon $\spadesuit U$ which has the property that $\boldsymbol{D} \subset \spadesuit U$ and that all points of $\boldsymbol{D}$ lie in $\spadesuit U^o$ except the vertex $(1,1)$. We check this as follows. First we use the strong confinement test to show that $\gamma(\clubsuit Q) \subset \spadesuit U^o$ for all $\gamma \in \Gamma$ except the identity and $\boldsymbol{R}$. The two sets $\clubsuit Q$ and $\boldsymbol{R}(\clubsuit Q)$ are polygons, and we see directly that $\clubsuit Q - \{(1,1)\}$ and $\boldsymbol{R}(\clubsuit Q) - \{(1,1)\}$ lie in $\spadesuit U^o$. We list the vertices of $\spadesuit U$ in §21.5. The approximation is very tight in certain places where we need it, and otherwise loose.

LEMMA 12.4. *Let $p \in \boldsymbol{D} - \bigcup \boldsymbol{D}_j$ be any point not in the set $\boldsymbol{\Gamma}(\mathbf{s})$. Then any primary map h which is defined and finite at p maps p outside $\boldsymbol{D}$.*

Proof: Since $\boldsymbol{D}$ is $\boldsymbol{\Gamma}$ invariant, it suffices to show that there are primary maps h_1, h_2, h_3 so that h_j is defined on $\clubsuit S_j - \mathbf{s}$ and $h_j(\clubsuit S_j - \mathbf{s}) \subset \boldsymbol{R}^2 - \boldsymbol{D}$. We set

$$h_1 = \boldsymbol{H}_{031}, \qquad h_2 = \boldsymbol{H}_{141}, \qquad h_3 = \boldsymbol{H}_{111}. \tag{12.7}$$

We use the denominator test to check that h_3 is defined on $\clubsuit S_3$. For $j = 1, 2$ we use the denominator test coupled with the subtraction trick to check that h_j is defined and finite on $\clubsuit S_j - \mathbf{s}$. Let $\clubsuit S_j' = \clubsuit S_j - \mathbf{s}$. We use the exclusion test to check that $h_j(\clubsuit S_j') \cap \spadesuit U^o = \emptyset$. Hence $h_j(\clubsuit S_j')$ is disjoint from $\boldsymbol{D} - \{(1,1)\}$. We also use the exclusion test to check that $h_j(\clubsuit S_j')$ is disjoint from the square of radius 1 centered at $(1,1)$. Hence $h_j(\clubsuit S_j') \cap \boldsymbol{D} = \emptyset$. $\square$

LEMMA 12.5. *$\boldsymbol{H}_{101}$ is good with respect to $\boldsymbol{D}_2$.*

Proof: Let $h = \boldsymbol{H}_{101}$.

(1) We use the denominator test to check that h is defined and finite on $\clubsuit T$.

(2) We check that $(h, \clubsuit T, 10, 45, 45)$ is expansive. Hence h is an expanding diffeomorphism on $\clubsuit T$.
(3) We check that $f(\clubsuit T) \cap (\boldsymbol{T} - \boldsymbol{C}_*) = \emptyset$ for all 10 primary elements f. Hence $h(\boldsymbol{D}_2) \subset \boldsymbol{C}_*$.
(4) Three of the edges of $\clubsuit T$ lie in the $\boldsymbol{D} - \bigcup \boldsymbol{D}_j$. By Lemma 12.4 and symmetry, h maps these three edges outside of $\boldsymbol{D}$.
(5) We use the exclusion test to check that $h(e) \cap \spadesuit U^o = \emptyset$ for the remaining non-diagonal edge e of $\clubsuit T$.
(6) By symmetry, h maps the diagonal edge of $\clubsuit T$ into the diagonal.

In summary, the restriction of h to $\clubsuit T$ is a diffeomorphism which maps the diagonal edge of $\clubsuit T$ into the diagonal and all remaining edges of $\clubsuit T$ disjointly from $\boldsymbol{D} - \{(1,1)\}$. From this, and bilateral symmetry, $\boldsymbol{D} \subset h(\boldsymbol{D}_2)$. $\square$

The multilayered Figure 12.7 shows an accurate picture of the previous lemma. On the left hand side, the small white shape is $\clubsuit T$. The dark grey shape is the half of $\boldsymbol{D}$ below the diagonal. The large medium grey shape is $h(\clubsuit T)$. It is hidden behind the dark grey piece. The right hand side of the figure shows a closeup. The reader can see much better pictures using the computer program.

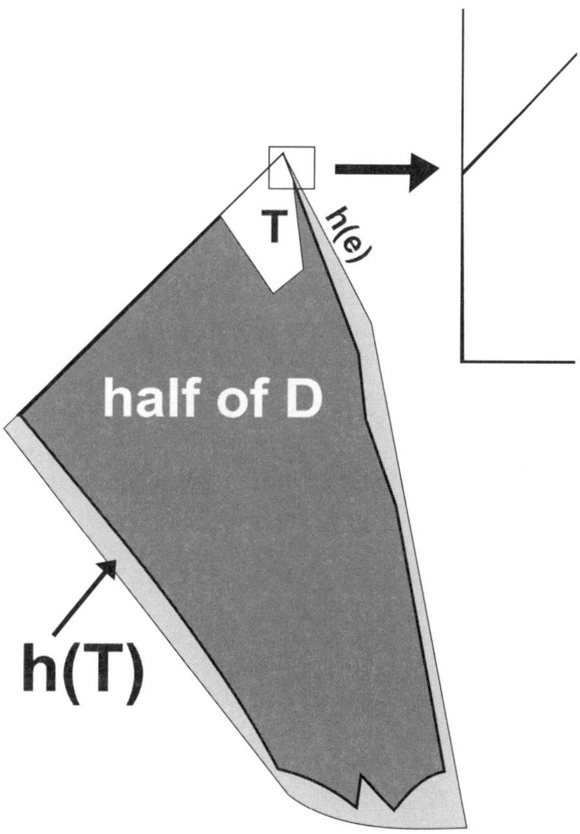

FIGURE 12.7. $\clubsuit T$ and $h(\clubsuit T)$

12.5. The Main Argument

Let $\boldsymbol{H} = \boldsymbol{B}H\boldsymbol{B}^{-1}$, where H is the projective heat map. Define

$$\boldsymbol{K}_k = \boldsymbol{H}^{-1}(\boldsymbol{D}) \cap \boldsymbol{D}_k. \tag{12.8}$$

Since $\boldsymbol{H}$ is a primary map, the restriction of $\boldsymbol{H}$ to $\boldsymbol{D}_k$ is a diffeomorphism onto its image. Hence $\boldsymbol{K}_k$ is a piecewise analytic disk for each $k = 1, \ldots, 6$. We use $\boldsymbol{K}_k$ in place of $\boldsymbol{D}_k$ because we have more control over the image: $\boldsymbol{H}(\boldsymbol{K}_k) = \boldsymbol{D}$.

We let $\heartsuit\boldsymbol{K}$ denote the subset of points $p \in \boldsymbol{K}_1 \cup \cdots \cup \boldsymbol{K}_6$ such that $\boldsymbol{H}^n(p) \in \boldsymbol{D}$ for all n. The set $\heartsuit\boldsymbol{K}$ is evidently $\boldsymbol{H}$-invariant.

Let μ denote the planar measure on $\boldsymbol{D}$ which is $\boldsymbol{\Gamma}$-invariant and agrees with Lebesgue measure on $\boldsymbol{D} \cap \boldsymbol{T}$. The measure μ is absolutely continuous with respect to Lebesgue measure. Below we will prove

LEMMA 12.6 (Measure Expansion). *For each $k = 1, \ldots, 6$, the restriction of $\boldsymbol{H}$ to $\boldsymbol{K}_k$ expands μ by a factor of at least 7.*

Let $\boldsymbol{K}(n)$ denote the set of points $p \in \heartsuit\boldsymbol{K}$ such that $\boldsymbol{H}^n(p) \in \boldsymbol{D}$. As discussed in §6.2, and in view of the results proved in the previous section, $\boldsymbol{K}(n)$ consists of 6^n topological disks. In view of the Measure Expansion Lemma, each of these disks has μ measure $O(7^{-n})$. Hence $\boldsymbol{K}(n)$ has μ measure $O(6^n/7^n)$, a quantity which tends to 0 with n. Being the nested intersection of sets of arbitrarily small μ measure, we see that $\heartsuit\boldsymbol{K}$ has μ measure 0. Since μ is absolutely continuous with respect to Lebesgue measure, we see that $\heartsuit\boldsymbol{K}$ has Lebesgue measure 0 as well.

Our next result refers to the expansion hypotheses discussed in §6.2.3.

LEMMA 12.7 (Metric Expansion). *There is some $\eta > 1$ and some tangent bundle function ρ on $\boldsymbol{D}$ such that the restriction of $\boldsymbol{H}$ to any disk of $\boldsymbol{K}(2)$ expands ρ by at least a factor of η.*

We have established the same hypotheses as in §6.2. We conclude from the construction there that $\heartsuit\boldsymbol{K}$ is a Cantor set and the restriction of $\boldsymbol{H}$ to $\heartsuit\boldsymbol{K}$ is conjugate to the 1-sided shift on 6 symbols.

Remark: It might have been nicer to be able to just work with the Euclidean metric for the results above, but I couldn't quite manage it. The underlying difficulty is that $\boldsymbol{\Gamma}$ does not act isometrically or by area-preserving maps on $\boldsymbol{D}$. Actually, the funny measures and metrics I use make the proofs easier. In the end, we just make Euclidean measurements, but we use the funny measure and metric to maximally exploit symmetry.

It only remains to pull everything back to $\mathcal{P}$ using the change of coordinates map $\boldsymbol{B}$. We define $\mathcal{JC} = \boldsymbol{B}^{-1}(\heartsuit\boldsymbol{K})$. Everything in sight takes place inside the subset of the configuration space consisting of star-shaped classes, and $\boldsymbol{B}$ is a diffeomorphism there which conjugates H to $\boldsymbol{H}$. Therefore, $\mathcal{JC}$ and $\heartsuit\boldsymbol{K}$ are homeomorphic via an ambient diffeomorphism, and the action of H on $\mathcal{JC}$ is conjugate to the action of $\boldsymbol{H}$ on $\heartsuit\boldsymbol{K}$. Hence $\mathcal{JC}$ is a H-invariant Cantor set of measure 0 and the restriction of H to $\mathcal{JC}$ is conjugate to the 1-sided shift on 6 symbols. This completes the proof of Theorem 1.5 modulo the expansion results mentioned above.

The rest of the chapter is devoted to proving the expansion results.

12.6. Proof of the Measure Expansion Lemma

LEMMA 12.8. *$\boldsymbol{H}$ expands μ by a factor of at least 7 when restricted to $\boldsymbol{K}_1$.*

Proof: Since μ is Γ-invariant, it suffices to prove this result for the primary map of our choosing. We choose $\boldsymbol{H}_{111}$. Since $\boldsymbol{K}_1 = \boldsymbol{\Gamma}(\clubsuit R)$, it suffices by symmetry to prove our result for the restriction of $\boldsymbol{H}_{111}$ to $\clubsuit R$. We have $\boldsymbol{H}_{111}(\clubsuit R) \subset \boldsymbol{T}$, as above. Hence

$$\boldsymbol{H}_{111}(\clubsuit R \cap \boldsymbol{K}_{10}) \subset \clubsuit Q. \tag{12.9}$$

So, we can simply make our analysis using Lebesgue measure. We use the area test to show that $\boldsymbol{H}_{111}$ satisfies $|\det d\boldsymbol{H}_{111}| \geq 7$ on $\clubsuit R$. □

LEMMA 12.9. *For each $k = 2,3,4,5,6$, the restriction of $\boldsymbol{H}$ to $\boldsymbol{D}_k$ expands μ by at least a factor of 7.*

Proof: By symmetry, it suffices to prove that

$$\boldsymbol{H} : \clubsuit T \cap \boldsymbol{K}_2 \to \boldsymbol{D}$$

expands μ by a factor of at least 7. In the region $\clubsuit T \cap \boldsymbol{K}_2$ we have $\lambda = \mu$, so it looks like we are on track to use the same symmetry as in the previous lemma. However, this time our luck runs out because the image of this set under $\boldsymbol{H}$ does not lie in the region of $\boldsymbol{D}$ which agrees with λ.

We fix the problem by proving that the restriction of *every* primary map to $\clubsuit T \cap \boldsymbol{K}_2$ expands λ by at least 7. Since $\lambda = \mu$ on a fundamental domain for the action of $\boldsymbol{\Gamma}$ on $\boldsymbol{D}$, and since μ is $\boldsymbol{\Gamma}$ invariant, this last result implies that the restriction of every primary map, including $\boldsymbol{H}$, to $\clubsuit T \cap \boldsymbol{K}_2$ also expands μ by a factor of 7.

Let X be any of the pieces of $\clubsuit T$ and let h be any primary element. We check that (X, h) is clean for every piece X of $\clubsuit T$ and every primary h. Next, we use the strong area test for each pair and show that $|\det\ dh| > 7$ on X. □

12.7. Proof of the Metric Expansion Lemma

First we need to define the function ρ on the tangent bundle of $\boldsymbol{D}$. We would like to just take the $\boldsymbol{\Gamma}$ invariant Riemannian metric which agrees with the Euclidean metric in the fundamental domain $\boldsymbol{D} \cap \boldsymbol{T}$. However, this does not really make sense because there are problems with piecing the metric together across the boundaries. We have to broaden our definitions to make sense of the construction we want.

There is a canonical fundamental domain $\boldsymbol{FD}$ for the action of $\boldsymbol{\Gamma}$ on the tangent bundle; it is the lift of the fundamental domain $\boldsymbol{D} \cap \boldsymbol{T}$ for the action of $\boldsymbol{\Gamma}$ on $\boldsymbol{D}$. A point (p, V) lies in $\boldsymbol{FD}$ iff either p lies in the interior of $\boldsymbol{D} \cap \boldsymbol{T}$ or else p lies on one of the two edges $e_0 = \partial_0 \boldsymbol{T} \cap \boldsymbol{D}$ or $e_2 = \partial_2 \boldsymbol{T} \cap \boldsymbol{D}$ and does not point out of $\boldsymbol{D} \cap \boldsymbol{T}$. Note that $\boldsymbol{FD}$ is, so to speak, an orbifold with mirrors. There are elements γ_0 and γ_2 which fix e_0 and e_2 pointwise. These elements act on the tangent bundle so as to pointwise fix the corresponding boundary components of $\boldsymbol{FT}$.

We let ρ be $\boldsymbol{\Gamma}$-invariant function on the tangent bundle which agrees with the function induced by the Euclidean norm on $\boldsymbol{FT}$. This definition is piecewise smooth. Thanks to the mirrors, this definition is everywhere continuous as well.

LEMMA 12.10. *$\boldsymbol{H}$ is ρ expanding on $\boldsymbol{K}_1$.*

Proof: Similar to Lemma 12.8, it suffices to show that $\boldsymbol{H}_{111}$ expands the Euclidean metric on $\clubsuit R$. We have already seen this in the proof of Lemma 12.3. □

LEMMA 12.11. *$\clubsuit T_4$ and $\clubsuit T_5$ are disjoint from $\boldsymbol{K}(2)$.*

Proof: See Figure 12.4 for a picture of $\clubsuit T_4$ and $\clubsuit T_5$. We first check using the denominator test that all the primary maps are defined and finite on $\clubsuit T_4$ and $\clubsuit T_5$. Using the exclusion test, we check for all $i \in \{4,5\}$ and all $j \in \{1,2,3\}$ and all primary h that $h(\clubsuit T_i) \cap \clubsuit T_j = \emptyset$. Likewise, we check that $h(\clubsuit T_i) \cap \clubsuit R = \emptyset$. This shows that $\boldsymbol{H}(\clubsuit T_4)$ and $\boldsymbol{H}(\clubsuit T_5)$ are disjoint from $\boldsymbol{D}_1 \cup \cdots \cup \boldsymbol{D}_6$. □

Note $\boldsymbol{K}_2 \cup \cdots \cup \boldsymbol{K}_6$ is the $\boldsymbol{\Gamma}$ orbit of $\clubsuit T$. So, to finish the proof of the Metric Expansion Lemma, it suffices to prove that $\boldsymbol{H}$ is locally ρ-expanding on $\clubsuit T_j \cap \boldsymbol{K}(2)$ for $j = 1,2,3$. We can ignore $\clubsuit T_4$ and $\clubsuit T_5$ by the previous result.

LEMMA 12.12. *$\boldsymbol{H}$ is ρ-expanding on $\clubsuit T_j \cap \boldsymbol{K}(2)$ for $j = 1,2$.*

Proof: We consider $\clubsuit T_1$ first. By symmetry it suffices to prove that $\boldsymbol{H}_{111}$ is locally ρ-expanding on $\clubsuit T_1 \cap \boldsymbol{K}(2)$. We use the exclusion test to show that $h(\clubsuit T_1)$ is disjoint from $\clubsuit T$ as long as h is a primary map other than H_{111} and RH_{111}.

$$\boldsymbol{H}_{111}(\clubsuit T_1 \cap \boldsymbol{K}(2)) \subset \boldsymbol{K}_2. \tag{12.10}$$

Note that ρ equals the Euclidean metric on $\boldsymbol{K}_2$. To finish our proof, we just have to show that $\boldsymbol{H}_{111}$ is Euclidean expanding on $\clubsuit T_1$. We check that

$$(\clubsuit T_1, \boldsymbol{H}_{111}, 15+\phi, 45, 220)$$

is expansive. Thus $\boldsymbol{H}_{111}$ is Euclidean expanding by a factor of at least

$$\frac{15+\phi}{\sqrt{45+220}} \approx 1.02$$

on $\clubsuit T_1$. We make for same checks for $\clubsuit T_2$ as we did for $\clubsuit T_1$. This time

$$(\clubsuit T_2, \boldsymbol{H}_{031}, 28, 350, 90)$$

is expansive. This H_{031} is Euclidean expanding by a factor of at least

$$\frac{28}{\sqrt{350+90}} \approx 1.33$$

on $\clubsuit T_2$. □

LEMMA 12.13. *$\boldsymbol{H}$ is locally ρ-expanding on $\clubsuit S_3 \cap \boldsymbol{K}(2)$.*

Proof: We would like to proceed as in the previous two cases, but this time our luck runs out and a result like equation (12.10) does not hold. We fix the problem by proceeding as in Lemma 12.9. We check that the following quintuples are expansive:

$$(\clubsuit S_3, \boldsymbol{H}_{001}, 20, 195, 195), \quad (\clubsuit S_3, \boldsymbol{H}_{011}, 20, 95, 295), \quad (\clubsuit S_3, \boldsymbol{H}_{021}, 24, 300, 270),$$
$$(\clubsuit S_3, \boldsymbol{H}_{031}, 20, 195, 195), \quad (\clubsuit S_3, \boldsymbol{H}_{041}, 20, 195, 195).$$

Since $\boldsymbol{R} \in \boldsymbol{\Gamma}$ is a Euclidean isometry, the checks above show that every primary map is Euclidean expanding on $\clubsuit S_3$. The closest call gives an expansion factor of at least

$$\frac{24}{\sqrt{300+270}} \approx 1.005.$$

By the same symmetry as in Lemma 12.9, we conclude that each primary element is ρ-expanding on $\clubsuit S_3 \cap \boldsymbol{K}(1)$. But then, *a fortiori*, the map $\boldsymbol{H}$ is ρ-expanding on $\clubsuit S_3 \cap \boldsymbol{K}(2)$. $\square$

This completes our proof of the Metric Expansion Lemma. Our proof of Theorem 1.5 is done.

12.8. Discussion

Here we discuss why our partition looks like it does. The triangle $\clubsuit R$ and the pentagon $\clubsuit T$ are the fundamental pieces used to define our 6 disks. The piece $\clubsuit S$ is a buffer between the triangle and the pentagon. The reason why we have 3 $\clubsuit S$-pieces rather than 1 is that the union of the $\clubsuit S$-pieces is highly nonconvex and our tests run much better on convex polygons.

We partitioned $\clubsuit T$ into smaller pieces because our methods are not sharp enough to treat $\clubsuit T$ as a single piece. It is true that we proved above that some primary map is Euclidean expanding on $\clubsuit T$. However, what we would really need to prove is that all primary maps are Euclidean expanding on $\clubsuit T$. Our methods could not get this. So, we partitioned $\clubsuit T$ into 5 smaller pieces. We couldn't get the expansion result on $\clubsuit T_4$ and $\clubsuit T_5$ but they turn out to be disjoint from the Cantor set. We found this partition of $\clubsuit T$ after quite a bit of trial and error.

All this careful partitioning, as well as the very close estimates on the expansion constants, is the price I pay for using the crude Expansion Test defined in §11.5. What would I have done if I couldn't jiggle the constants to make the lemmas in the preceding section succeed? The answer is that I would have enhanced the Expansion test so that it could measure expansion not just with respect to the Euclidean metric but also with respect to any suitably defined inner product. The easiest extension would be to allow inner products represented by rational matrices.

We would have gotten better results if we instead used the inner product with respect to which the differential action of $\boldsymbol{\Gamma}$ at the regular class be an action by similarities. Call this the *adapted inner product*. The auxiliary linear transformation T used to create Figure 1.4 carries the adapted inner product to the dot product. The problem with using the adapted inner product is that it is defined over a quartic number field rather than $\boldsymbol{Q}$, so I would have needed to beef up my calculations to handle it. One work-around would be to instead use a rational approximation to the adapted quadratic form. Perhaps this approach would have saved the trouble of subdividing $\clubsuit T$ into smaller pieces. In the end, I chose to work carefully with the partition rather than complicate the Expansion Test.

There is one more point I want to discuss. Referring to Figure 12.6, we think of $\spadesuit U \cup \boldsymbol{R}(\spadesuit U)$ as a polygonal approximation to $\boldsymbol{D}$. It might have been simpler to approximate $\boldsymbol{D}$ with a single convex polygon, but we couldn't find one which was close enough to $\boldsymbol{D}$ to be useful.

CHAPTER 13

Towards the Quasi Horseshoe

13.1. The Target

This chapter contains the first half of the proof of Theorem 1.6. In the end, we want to show that all points of $\mathcal{J} - \mathcal{J}C$ with well-defined orbits land in a certain Cantor band $\mathcal{J}A$. The set $\mathcal{J}A$ is contained in the set

$$\boldsymbol{B}^{-1}(\boldsymbol{\Gamma}(\clubsuit P)), \tag{13.1}$$

Where $\clubsuit P$ is the union of 8 quadrilaterals shown in Figure 13.1. We list the vertices of these quadrilaterals in §21.6.1. When we restrict suitable choices of the primary maps to $\clubsuit P$ (and make a tiny modification) we get a quasi-horseshoe.

For comparison, Figure 13.1 also shows the set $\clubsuit Q$ and also the lightly shaded set of points representing star convex pentagon classes. We also show 4 special points $\boldsymbol{p}$, $\boldsymbol{q}$, $\boldsymbol{r}$, and $(1,1)$ on $\boldsymbol{T}$. The point $\boldsymbol{p}$ is the top vertex of $\boldsymbol{T}$ and the points $\boldsymbol{q}$ and $\boldsymbol{r}$ turn out to be cone points in the sense of Theorem 1.8. The Cantor bands we eventually find in $\clubsuit P$ continue leftward and then pinch down to $\boldsymbol{q}$ and $\boldsymbol{r}$. We list coordinates for these special points in §21.4.

Definition: Let $q \in \boldsymbol{R}^2$ be some point.

(1) q is *pre-convex* if there is some $h \in \mathbf{S}$ such that $\boldsymbol{B}^{-1}h(q) \in \boldsymbol{C}$.
(2) q is *pre-*$\clubsuit P$ if there is some $h \in \mathbf{S}$ such that $h(p) \in \clubsuit P$.
(3) q is *pre-bad* if $p \in \Lambda$ and $\boldsymbol{H}^n(p) \in \{\mathbf{p}, \mathbf{q}\}$ for some n.

Remark: Condition 3 is different from Conditions 1 and 2 because Condition 3 defines a specific countable set of points on the diagonal boundary of $\boldsymbol{T}$.

Here is the goal of this chapter.

THEOREM 13.1. *Each point of $\boldsymbol{T} - \boldsymbol{D}_0$ is pre-convex, pre-$\clubsuit P$, or pre-bad.*

13.2. The Outer Layer

Figure 13.2 shows 8 polygons

$$\clubsuit A, \clubsuit B, \spadesuit C, \spadesuit D, \spadesuit E, \clubsuit F, \clubsuit G, \clubsuit H.$$

The polygons $\spadesuit C$, $\spadesuit D$, $\spadesuit E$ lie outside the fundamental domain. We only show the intersection of $\spadesuit D$ with the picture window. The whole piece $\spadesuit D$ is enormous; the whole tile is fairly close in size and shape to the square $[-250, 0]^2$. We list the vertices in §21.6.2. The white dot on the right boundary of $\clubsuit G$ will have significance, much later, in the proof of Lemma 17.3.

LEMMA 13.2. *All the points in the 8 above polygons are preconvex except $\boldsymbol{p}, \boldsymbol{q}, \boldsymbol{r}$.*

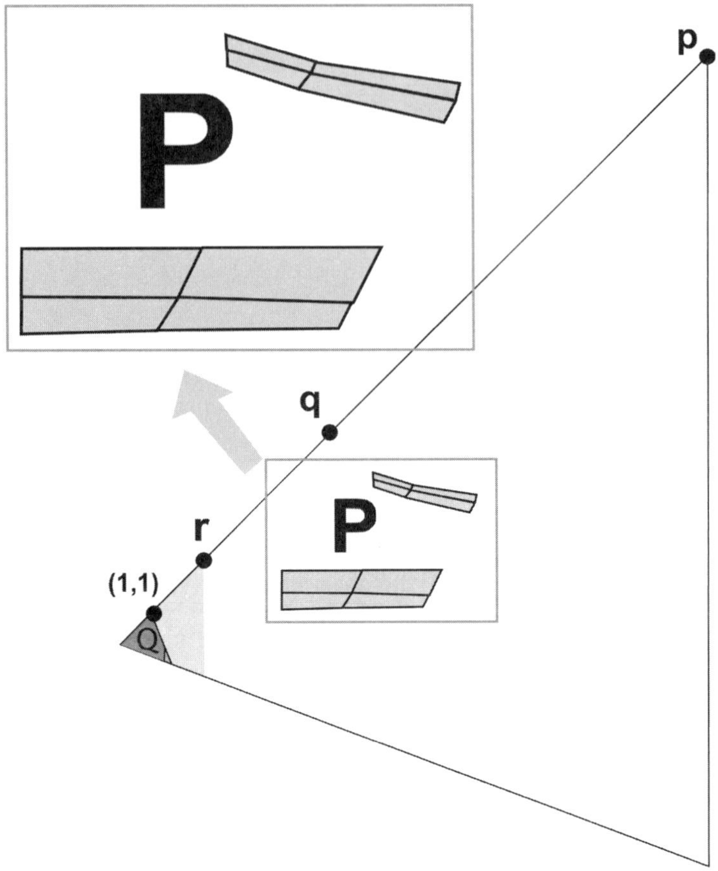

FIGURE 13.1. $\clubsuit P$ sitting inside $\boldsymbol{T}$

Proof: Let $\Theta = \boldsymbol{B}^{-1}\boldsymbol{H}$, as in equation (10.6). Our proof here repeatedly uses Corollary 9.2. Corollary 9.2, when combined with the definition of *preconvexity*, has the following consequence: If $\Theta(p) \in (0,1)^2$ then p is pre-convex.

- We use the denominator test and the variation trick to check that Θ is defined and finite on $\clubsuit A - \boldsymbol{p} - \boldsymbol{q}$.

- We use the denominator test and the subtraction trick to check that Θ is defined and finite on $\clubsuit G - \boldsymbol{q} - \boldsymbol{r}$ and $\clubsuit H - \boldsymbol{q}$.

- We use the (plain) denominator test to check that Θ is defined and finite on the remaining pieces.

We use the strong variant of the confinement test to check that

$$\Theta(\clubsuit B \cup \spadesuit C),\ \Theta(\spadesuit D) \subset (0,1)^2. \tag{13.2}$$

We use the (weak) confinement test to check that $\Theta(\clubsuit A - \boldsymbol{p} - \boldsymbol{q})$ is contained in the square of side length $1 - 2\phi^{-4}$ centered at $(1/2, 1/2)$. Hence

$$\Theta(\clubsuit A - \boldsymbol{p} - \boldsymbol{q}) \subset (0,1)^2. \tag{13.3}$$

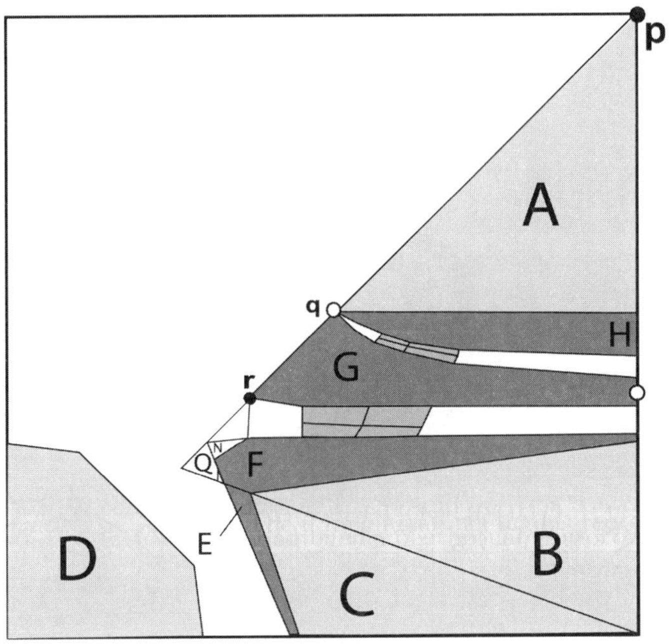

FIGURE 13.2. The outer layer

We now use the confinement test to check that

$$\boldsymbol{H}_{031}(\spadesuit E),\ \boldsymbol{H}_{141}(\clubsuit F),\ \boldsymbol{H}_{021}(\clubsuit G - \boldsymbol{q} - \boldsymbol{r}),\ \boldsymbol{H}_{131}(\clubsuit H - \boldsymbol{q}) \subset \spadesuit D. \tag{13.4}$$

This completes the proof. □

Remark: We will not have to use the strong version of the confinement test again. So, the remaining confinement tests use the WPDA, without any tricks, as described in the previous chapter. Also, the exclusion test uses the WPDA without any tricks. We mention this so that the reader knows exactly which tests we are performing.

13.3. The Inner Layer

Now we will finally encounter some pre-$\clubsuit P$ points. Figure 13.5 shows the collections of pieces labeled $\clubsuit I, \clubsuit J, \clubsuit K, \clubsuit L$. This time each letter stands for several pieces, according to the following scheme. There are essentially 4 rows of pieces, and each tile gets an index according to the row it sits in. The top row gets index 1 and the next row gets index 2, etc. Thus, the four $\clubsuit J$ pieces are $\clubsuit J_1, \clubsuit J_2, \clubsuit J_3, \clubsuit J_4$. We list the coordinates of these pieces in §21.6.3. These pieces flank the main set of interest, $\clubsuit P$. Figure 13.3 also shows $\clubsuit P$. It is the union of the 8 lightly shaded pieces.

LEMMA 13.3. *All points of $\clubsuit I \cup \clubsuit J \cup \clubsuit K \cup \clubsuit L$. other than* **q** *and* **r** *are either pre-convex or pre-$\clubsuit P$.*

Proof: We define the union of the previous pieces: $\clubsuit X = \clubsuit A \cup \cdots \cup \clubsuit H$. We also define the primary elements:

$$h_1 = \boldsymbol{H}_{111}, \qquad h_2 = \boldsymbol{H}_{041}, \qquad h_3 = \boldsymbol{H}_{141}, \qquad h_4 = \boldsymbol{H}_{021}. \tag{13.5}$$

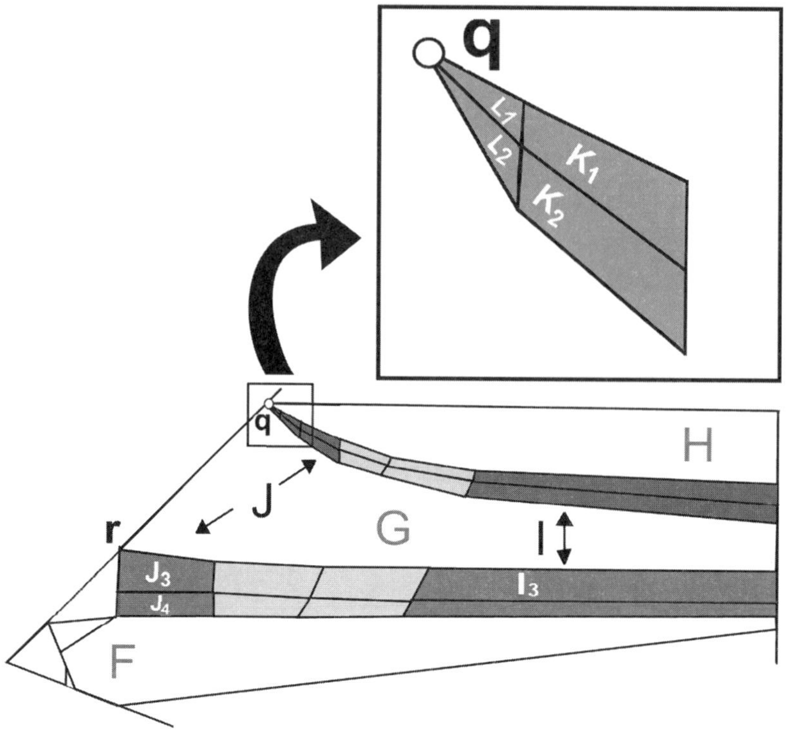

FIGURE 13.3. The inner layer

Finally, we define auxiliary polygons $♠Y_1$ and $♠Y_2$ which have the following properties:

$$♠Y_1 \subset ♣P \cup ♣X, \qquad ♠Y_2 \subset ♣P \cup ♣X \cup ♣I. \tag{13.6}$$

These polygons are taken to be nearly as large as possible. We list the vertices for these polygons in §21.6.3. Figure 13.4 shows a fairly accurate picture.

For $k = 1, 2, 3, 4$ we check that h_k is defined on $♣I_k$. Then we use the confinement test to show

$$h_k(♣I_k) \subset ♠Y_1. \tag{13.7}$$

Next, letting $♣V$ stand for any of $♣J, ♣K, ♣L$, we check that h_k is defined and finite on $♣V_k - \boldsymbol{q} - \boldsymbol{r}$ and that

$$h_k(♣V_k - \boldsymbol{q} - \boldsymbol{r}) \subset ♠Y_2. \tag{13.8}$$

By construction, each point of $♠Y_1$ is either preconvex or pre-$♣P$. Hence, each point of $♣I$ is either preconvex or pre-$♣P$. But then each point of $♠Y_2$ is either preconvex or pre-$♣P$. Hence, each point of $♣J, ♣K, ♣L$, except $\boldsymbol{q}$ and $\boldsymbol{r}$, is either preconvex or pre-$♣P$. □

Figure 13.4 shows a pretty accurate drawing of the pieces $♠Y_1$ and $♠Y_2$. Note that the piece $♠Y_2$ continues behind the darker pieces on the right hand side of Figure 13.4.

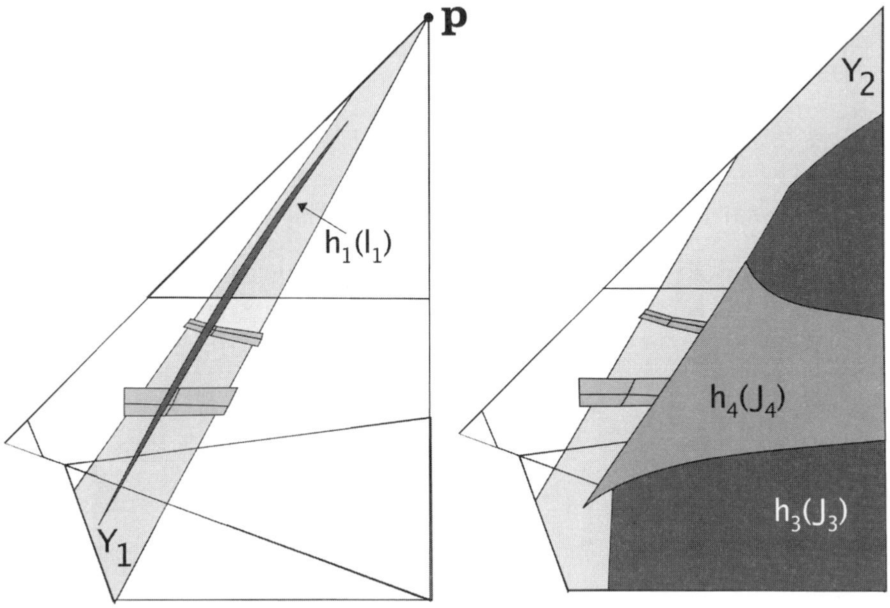

FIGURE 13.4. ♠Y_1 and ♠Y_2 and some representative images

13.4. The Last Three pieces

To finish the proof of Lemma 13.1 we need to deal with 3 exceptional pieces. Figure 13.5 shows the pieces ♣M, ♣N, ♣O, ♣Q. We are not interested in ♣Q but we draw it to illustrate how the other pieces fit around it. Figure 13.5 also indicates the locations of some of the pieces we have already considered, to help give a better picture for how these last pieces fit in. We list the coordinates of the vertices in §21.6.4.

LEMMA 13.4. *All points of ♣M are pre-convex.*

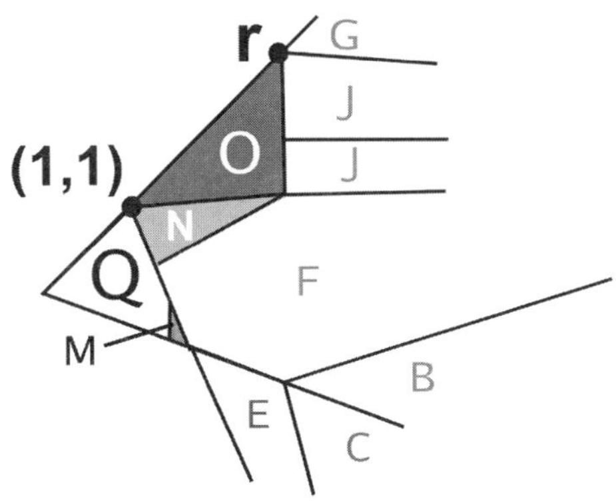

FIGURE 13.5. The pieces ♣M, ♣N, ♣O, ♣Q

Proof: We introduce a quadrilateral ♠Y_3 which is contained in ♣$A \cup$♣$H - \boldsymbol{p} - \boldsymbol{q}$. We check that $\boldsymbol{H}_{111}$ is defined and finite on ♣M and we check that $\boldsymbol{H}_{111}$(♣M) $\subset$ ♠Y_3. We list the coordinates of ♠Y_3 in §21.6.3. □

The rest of our proofs use the element $h = \boldsymbol{H}_{101}$. Note that $(1,1)$ is a fixed point of h. This makes the rest of our arguments more subtle.

LEMMA 13.5. *All points of* ♣N *except* $(1,1)$ *are pre-convex.*

Proof: The union ♠Y = ♣$N \cup$ ♣$B \cup$ ♣$F \cup$ ♠$C \cup$ ♠E is a convex polygon and obviously contains ♣N. We use the confinement test to show that h(♣N) $\subset$ ♠Y. We then check that the quintuple (♣$N, h, 36, 240, 110$) is expansive. Let $p \in$ ♣$N - \{(1,1)\}$. Since h fixes the vertex $(1,1)$ and is an expanding map, the infinite sequence $\{h^k(p)\}$ cannot remain in ♣N forever. Hence, there is some k such that $h^k(p) \subset$ ♠$Y -$ ♣N. But all points in this latter set are pre-convex. □

Now we come to the last piece ♣O. This one is the most interesting.

LEMMA 13.6. h(♣O) $\subset (\boldsymbol{T} - \boldsymbol{D}_0) \cup \{(1,1)\}$. *Moreover, h is an expanding map on* ♣O.

Proof: Let ♠Y be the triangle whose vertices are $(1,1)$ and the two vertices of $\boldsymbol{T}$ on the vertical edge. We have ♠$Y \subset \boldsymbol{T} - \boldsymbol{D}_0 \cup \{(1,1)\}$. We use the confinement test to show that h(♣O) $\subset Y_P$. Finally, we check that (♣$O, h, 43, 240, 200$) is expansive. □

Now we show that every point of ♣O is either preconvex pre-♣P or pre-bad. Let $p \in$ ♣O. We check that all the primary maps blow up at $\boldsymbol{q}$ and $\boldsymbol{p}$ and moreover that $h(\boldsymbol{r}) = \boldsymbol{p}$. So, if some power of h maps p into $\boldsymbol{q} \cup \boldsymbol{r}$, then p is pre-bad. Otherwise, thanks to the expanding nature of h, there is some k such that h^k lies in one of the other partition pieces of $\boldsymbol{T} - \boldsymbol{D}_0$ and is either preconvex or pre-♣P by one of the results above.

We have now dealt with all the partition pieces. This proves Theorem 13.1.

Remark: There is a little more we want to say. h maps the diagonal edge of ♣O into the diagonal edge of $\boldsymbol{T}$. For this reason, the set of points which are mapped into $\boldsymbol{q} \cup \boldsymbol{r}$ by powers of h – in other words, the full pre-image of $\boldsymbol{q} \cup \boldsymbol{r}$ under powers of h – is just a countable collection on the diagonal edge of $\boldsymbol{T}$ between $(1,1)$ and $\boldsymbol{r}$, which accumulates on $(1,1)$. These points turn out to play an important role in the proof of Theorem 1.10.

CHAPTER 14

The Quasi Horseshoe

14.1. Overview

We continue with the notation from the previous chapter. $\clubsuit P = \{\clubsuit P_i^j\}$ denotes the union of 8 special quadrilaterals. The upper index $j = 1, 2$ indexes quads going from left to right. The lower index $i = 1, 2, 3, 4$ indexes quads going from top to bottom. Each $\clubsuit P_i^j$ is an adapted quadrilateral in the sense of §6.5. Let

$$\clubsuit P^j = \clubsuit P_1^j \cup \clubsuit P_2^j \cup \clubsuit P_3^j \cup \clubsuit P_4^j. \tag{14.1}$$

So, $\clubsuit P^1$ and $\clubsuit P^2$ are the left and right halves of $\clubsuit P$. Let

$$\clubsuit P_j = \clubsuit P_j^1 \cup \clubsuit P_j^2 \tag{14.2}$$

For each j, these two sets share a timelike edge. Even though $\clubsuit P_j$ is a hexagon, it counts as a quad according to the definition in §6.5.

We define the following 4 primary elements:

$$h_1 = \boldsymbol{H}_{111}, \quad h_2 = \boldsymbol{H}_{041}, \quad h_3 = \boldsymbol{H}_{141}, \quad h_4 = \boldsymbol{H}_{021}. \tag{14.3}$$

We use the strong denominator and area tests to show that h_j is defined, finite, and a local diffeomorphism on $\clubsuit P_j$. We now (attempt to) define $F : \clubsuit P \to \boldsymbol{R}^2$ by saying that $F = h_j$ on $\clubsuit P_j$.

There is a technical problem. Since $\clubsuit P_1$ and $\clubsuit P_2$ share a common boundary, we do not know whether to define $F = h_1$ or $F = h_2$ there. The same goes for $\clubsuit P_3$ and $\clubsuit P_4$. However, we will prove below that h_1 and h_2 both map $\clubsuit P_1 \cap \clubsuit P_2$ into the union of pieces containing only pre-convex points. The same goes for h_3 and h_4 acting on $\clubsuit P_3 \cap \clubsuit P_4$. By continuity, the same holds for a thin tubular neighborhood of these arcs. We move the vertices of our quads inward by some tiny amount, say 10^{-100}, so that they are pairwise disjoint. We redefine $\clubsuit P$ to be the union of the modified quads. Now we can define $F : \clubsuit P \to \boldsymbol{R}^2$ unambiguously. We keep the definition of $\clubsuit P^1$ and $\clubsuit P^2$ the same as we want these two halves to fit perfectly with the rest of our tiling.

Here is the goal of the first part of the chapter.

THEOREM 14.1. *$F : \clubsuit P \to \boldsymbol{R}^2$ is a quasi-horseshoe.*

Following the proof of Theorem 14.1, we prove proving Theorem 1.6. The main difference between Theorem 1.6 and Theorem 14.1 is that Theorem 14.1 concerns a map defined piecewise by the action of 4 different elements of the semigroup $\boldsymbol{S}$ whereas Theorem 1.6 concerns the action of the single map H on the space $\mathcal{P}$.

14.2. Existence of The Quasi Horseshoe

Recall that $\vee$ is the standard light cone, defined in §6.5. We first check that F is quasi-hyperbolic on $\clubsuit P$.

LEMMA 14.2. *For any $p \in \clubsuit P$, the differential dh_p maps $\vee$ strictly inside itself and expands distances in $\vee$ by a factor of at least* 5.

Proof: For each of the 8 quadrilaterals comprising $\clubsuit P$, we check the first statement using the cone test and the second statement using the stretch test. See §11.8 and §11.9. □

Now we establish the interlacing property. We define

$$\clubsuit X = \clubsuit A \cup \clubsuit B \cup \clubsuit F \cup \clubsuit G \cup \spadesuit C - \boldsymbol{p} - \boldsymbol{q} - \boldsymbol{r}. \tag{14.4}$$

Figure 14.1 shows $\clubsuit P \cup \clubsuit X$ as well as auxiliary quadrilaterals $\spadesuit Y^1, \spadesuit Y^2, \spadesuit Z$. We list the vertices for these pieces in §21.6.5.

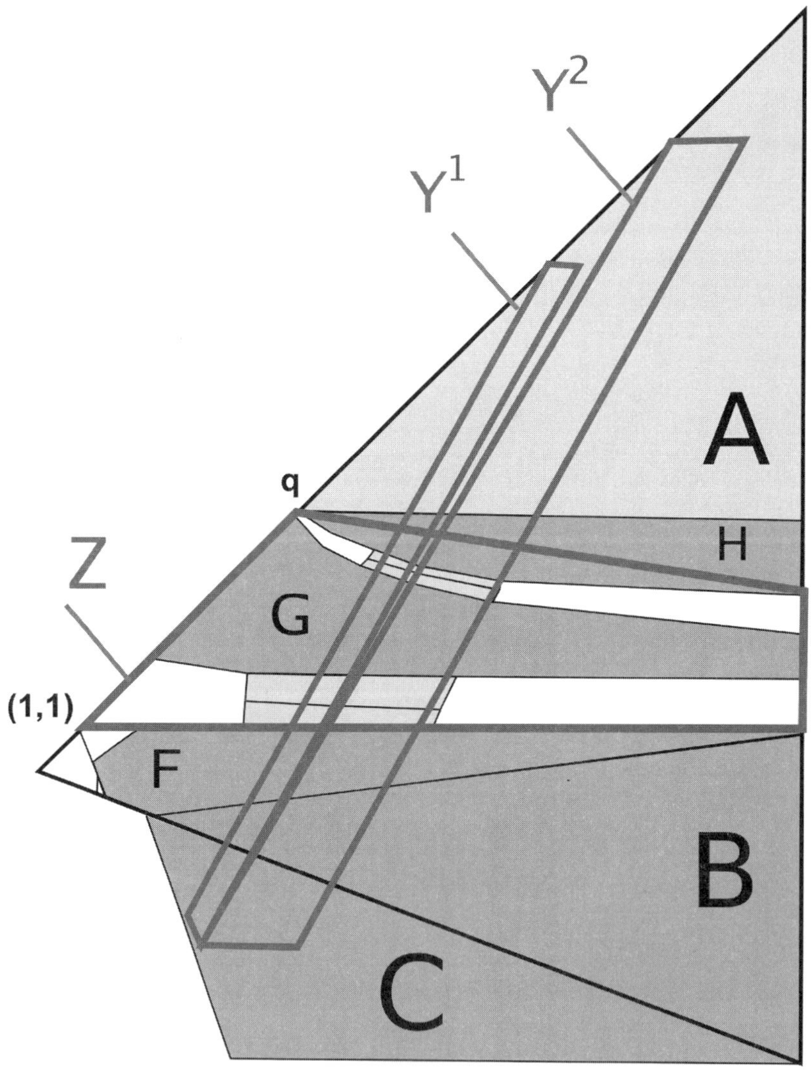

FIGURE 14.1. The sets $\clubsuit X$ and $\spadesuit Y^1$ and $\spadesuit Y^2$ and $\spadesuit Z$

Our analysis in the previous chapter shows that every point of $\clubsuit X$ is pre-convex. The reason we keep $\clubsuit P^1$ and $\clubsuit P^2$ unperturbed is that then these pieces fit together perfectly with $\clubsuit X$, forming a pair of bridges connecting the disconnected components of $\clubsuit X$.

LEMMA 14.3. *For $i=1,2$, we have $h_i(\clubsuit P_i) \subset \clubsuit P^1 \cup \clubsuit X$. Moreover, $h_i(\clubsuit P_i)$ interlaces $\clubsuit P^1_1, \clubsuit P^1_2, \clubsuit P^1_3, \clubsuit P^1_4$.*

Proof: For $j=1,2$, we use the confinement test to check that

$$h_i(\clubsuit P_i^j) \subset \spadesuit Y^1$$

for $i=1,2$ and $j=1,2$. Next we check that $\spadesuit Y^1 \subset \clubsuit P^1 \cup \clubsuit X$. (This is just a relation between polygons.) Finally, we use the exclusion test to check that h_i maps both horizontal edges of $\clubsuit P_i^j$ outside the quadrilateral $\spadesuit Z$, one above and on below. Since this works for $\clubsuit P_i^1$ and $\clubsuit P_i^2$, it works for their union, $\clubsuit P_i$. (The above/below choice can't switch between the two halves, by continuity.) □

LEMMA 14.4. *For $i=3,4$, we have $h_i(\clubsuit P_i) \subset \clubsuit P^2 \cup \clubsuit X$. Moreover, $h_i(\clubsuit P_i)$ interlaces $\clubsuit P^2_1, \clubsuit P^2_2, \clubsuit P^2_3, \clubsuit P^2_4$.*

Proof: The proof is as in Lemma 14.3 except that we use $\spadesuit Y^2$ in place of $\spadesuit Y^1$. □

This shows that $F : \clubsuit P \to \boldsymbol{R}^2$ is a quasi-horseshoe, and completes the proof of Theorem 14.1.

14.3. The Invariant Cantor Band

The rest of the chapter is devoted to proving Theorem 1.6. Here is the result stated again. In this result, $\mathcal{J}C$ is the Cantor set from Theorem 1.5.

THEOREM 14.5. *$\mathcal{J}$ contains a measure 0 forward-invariant Cantor band $\mathcal{J}A$ such that*

$$\mathcal{J} = \mathcal{J}C \cup \bigcup_{k=0}^{\infty} H^{-k}(\mathcal{J}A).$$

$\mathcal{J}A$ is an open subset of $\mathcal{J}$ in the subspace topology. The action of H in a neighborhood of $\mathcal{J}A$ is the 10-fold covering of a quasi-horseshoe.

In this section we define $\mathcal{J}A$ and show that it is a measure 0 Cantor band. Let $F : \clubsuit P \to \boldsymbol{R}^2$ be the quasi-horseshoe from Theorem 14.1. Let $\heartsuit A$ be the Cantor band guaranteed by the results in §6.5. Every point of $\clubsuit P - \heartsuit A$ is pre-convex and $\heartsuit A$ has measure 0.

The set $\clubsuit P$ lies in the interior of the fundamental domain $\boldsymbol{T}$ and above the line $y=\phi^2$. For this reason, the restriction $B^{-1} : \clubsuit P \to \boldsymbol{R}^2$ is a diffeomorphism. Hence $\boldsymbol{B}^{-1}$ is a diffeomorphism in a neighborhood of $\clubsuit P$.

Recall that $\mathcal{P}$ is the space of projective classes of pentagons. The Gauss group Γ acts on $\mathcal{P}$ as a group of diffeomorphisms. We define

$$\mathcal{J}A = \bigcup_{\gamma \in \Gamma} \gamma \circ \boldsymbol{B}^{-1}(\heartsuit A). \tag{14.5}$$

$\mathcal{J}A$ is the union of 10 pairwise disjoint Cantor bands of measure 0. Hence $\mathcal{J}A$ has measure 0.

14.4. Covering Property

In this section we show that the action of the projective heat map H in a neighborhood of $\mathcal{J}A$ is a 10-fold covering of a quasi-horseshoe. Mostly, the proof just amounts to saying what we actually mean. The set

$$\mathcal{U} = \Gamma \boldsymbol{B}^{-1}(\clubsuit P) = \boldsymbol{B}^{-1}\boldsymbol{\Gamma}\clubsuit P \tag{14.6}$$

is a Γ-equivariant neighborhood of $\mathcal{J}\mathcal{A}$ in $\mathcal{P}$. The map $\boldsymbol{B} : \mathcal{U} \to \boldsymbol{\Gamma}\clubsuit P$ is a diffeomorphism, by symmetry and by the remarks above about $\boldsymbol{B}$ being a diffeomorphism in a neighborhood of $\boldsymbol{B}^{-1}\clubsuit P$. The neighborhood $\mathcal{U}$ has 10 components, each diffeomorphic to the topological disk $\clubsuit P$. We let $\pi : \mathcal{U} \to \clubsuit P$ be the composition of $\boldsymbol{B}$ with the quotient map from $\boldsymbol{\Gamma}\clubsuit P$ to $\clubsuit P$.

By construction, π is a 10-fold covering map that carries each H orbit in $\mathcal{J}A$ to an F-orbit in $\heartsuit A$. This is the second point of Theorem 1.6.

Remark: The orbit-respecting property is also true through out all of $\mathcal{U}$, and not just on $\mathcal{J}\mathcal{A}$, but we lose the covering property as soon as the point exits $\Gamma\clubsuit P$.

14.5. Subspace Property

In this section we prove that $\mathcal{J}A$ is an open subset of $\mathcal{J}$ in the subspace topology. Let $\mathcal{N}$ be the space of nonconvex pentagon classes. We have $\mathcal{J} \subset \mathcal{N}$ by Theorem 1.3. Given $p \in \mathcal{N}$, there is a point $q \in \boldsymbol{T}$ such that $q = \boldsymbol{B}(\gamma(p))$ for some $\gamma \in \Gamma$. We have seen in §10 that $\boldsymbol{T}$ is a fundamental domain for the action of $\boldsymbol{\Gamma}$ on $\boldsymbol{B}(\mathcal{N})$. We call q an *associate* of p.

LEMMA 14.6. *No associate of a point in $\mathcal{J}$ is preconvex.*

Proof: Recall that $\mathcal{C}$ is the space of convex projective classes. Suppose $q \in \boldsymbol{T}$ is a preconvex associate of $p \in \mathcal{J}$. Then there are primary elements $h_1, \ldots, h_m$ so that

$$\boldsymbol{B}^{-1}h_m \ldots h_1(q) \in \mathcal{C}.$$

The map on the left side is defined and finite in a neighborhood U_q of q and maps this entire neighborhood into $\mathcal{C}$. Since $\boldsymbol{B}^{-1}$ is defined and finite at q, we can choose U_q small enough so that $\boldsymbol{B}^{-1}$ is also defined and finite on U_q.

Say that $p \in \boldsymbol{R}^2$ is N-*good* if every word of length N in $\boldsymbol{B}, \boldsymbol{B}^{-1}, G, R, H$ is well defined and finite on p. The set of N-good points is open dense for all N. Choose some constant N much larger than n and approximate q by an N-good point $q' \in U_q$. If we choose U_q small enough then $p' = \boldsymbol{B}^{-1}(q')$ is so close to one of the points in $\Gamma(p)$ that $H^n(p') \in \mathcal{N}$. Since q' is N-good, we have

$$\boldsymbol{B}^{-1}h_m \ldots h_1(q') = \gamma H^m(p')$$

for some $\gamma \in \Gamma$. But the left side is in $\mathcal{C}$ and the right side is in $\mathcal{N}$. This is a contradiction. $\square$

Choose an arbitrary $p \in \mathcal{J}A$. We just need to show that there is some small open set U of $\mathcal{P}$ such that $U \cap \mathcal{J} = \mathcal{U} \cap \mathcal{J}A$. By symmetry, we can take $p \in \boldsymbol{B}^{-1}(\heartsuit A)$ and $U \subset \boldsymbol{B}^{-1}(\clubsuit P)$. Let $q \in \mathcal{J} \cap \Delta$. The point $p = \boldsymbol{B}(q) \in \clubsuit P$ is associated to q. Hence p is not pre-convex, by Lemma 14.6. Since $q \in \clubsuit P$ and q is not pre-convex, we must have $q \in \heartsuit A$. Hence $p \in \mathcal{J}A$.

14.6. Attracting Property

To finish the proof of Theorem 1.6 we establish the following equation.

$$\mathcal{J} = \mathcal{J}C \cup \bigcup_{k=0}^{\infty} H^{-k}(\mathcal{J}A). \tag{14.7}$$

Here $\mathcal{J}C$ is the Cantor set from Theorem 1.5. Equation (14.7) is equivalent to the statement that every point in $\mathcal{J} - \mathcal{J}C$ is mapped into $\mathcal{J}A$ by some power of H.

This result relies on two technical results. The first technical result is similar in spirit to Lemma 14.6. In particular, it uses the notation of associate points defined in connection with Lemma 14.6. We say some words about the first of these lemmas in advance of stating the lemma, to help explain what it means. Suppose we have some subset $\mathcal{V} \subset \boldsymbol{R}^2$ and we want to decide when some high power of H maps a point of $\mathcal{J}$ into $\mathcal{V}$. We suppose that $\boldsymbol{B}$ is a diffeomorphism from an open neighborhood of $\mathcal{V}$ to an open neighborhood of $\boldsymbol{V}$ which carries $\mathcal{V}$ to $\boldsymbol{V}$. We want to transfer our knowledge of the action of the semigroup $\boldsymbol{S}$ on $\boldsymbol{V}$ back to $\mathcal{V}$. Our result uses the notation just established.

LEMMA 14.7. *Suppose $p \in \mathcal{J}$, and q is an associate of p. If $h(q) \in \boldsymbol{V}$ for some $h \in \boldsymbol{S}$, then there is some $\gamma \in \Gamma$ and some m such that $\gamma H^m(p) \in \mathcal{V}$.*

Proof: Since $\mathcal{J}$ is Γ-invariant, as is the notation of associates, we can assume without loss of generality that $q = \boldsymbol{B}(p)$. If $q \in \boldsymbol{V}$ there is nothing to prove. Consider the case when q does not lie in $\boldsymbol{V}$. In this case, h is some nontrivial composition of m primary elements $h_1, \ldots, h_m$. We have

$$h_m \ldots h_1(q) \in \boldsymbol{V}.$$

Here h is defined and finite in a neighborhood of q.

To streamline the proof, we write $a \sim a'$ if a and a' are $a, a' \in \boldsymbol{R}^2$ are extremely close. As in Lemma 14.6, we choose N much larger than m and let p' be an N-good point with $p \sim p'$. By continuity,

$$\boldsymbol{B}^{-1}h(q) \sim \boldsymbol{B}^{-1}h(q') = \gamma H^m(p') \sim \gamma H^m(p).$$

Letting $p' \to p$ we get $\boldsymbol{B}^{-1}(q) = \gamma H^m(p)$, the desired conclusion. □

Here is the first corollary of Lemma 14.7.

LEMMA 14.8. *The associate of a point in $\mathcal{J}$ is not pre-bad.*

Proof: Let $p \in \mathcal{J}$ be the point and let $q \in \boldsymbol{T}$ be an associate point. In Lemma 14.7 we take $\boldsymbol{V} = \mathbf{p} \cup \mathbf{q}$, the points from Theorem 13.1. We have

$$\boldsymbol{B}^{-1}(\mathbf{p}) = (1,1), \qquad \boldsymbol{B}^{-1}(\mathbf{q}) = (\psi,\psi), \qquad \psi = \frac{1+\sqrt{13}}{2}. \tag{14.8}$$

We take

$$\mathcal{V} = \{(1,1)\} \cup \{\psi,\psi\}.$$

By Lemma 14.7, we see that $\gamma H^m(p) \in \mathcal{V}$ for some $\gamma \in \Gamma$ and some m. But then $H^m(p)$ lies in the Γ orbit of $\mathcal{V}$. But no point in the Γ orbit of $\mathcal{V}$ has a well-defined orbit. Hence p does not have a well defined orbit. But all points in $\mathcal{J}$ have well-defined orbits, by definition. This is a contradiction. □

Now we establish equation (14.7). Let $p \in \mathcal{J} - \mathcal{J}C$. We know that p has some associate point $q \in \boldsymbol{T}$. There are 3 cases.

Case 1: Suppose $q \notin \boldsymbol{D}$. Theorem 13.1 tells us that q is either pre-convex or pre-$\clubsuit P$ or pre-bad. Lemma 14.6 says that q is not pre-convex. Lemma 14.7 says that q is not pre-bad. Hence q is pre-$\clubsuit P$. Taking $V = \clubsuit P$ in Lemma 14.7 we see that some forward iterate of p has an associate in $\clubsuit P$. So it suffices to assume that $q \in \clubsuit P$. Since q is not pre-convex, $q \in \heartsuit A$ and $p \in \mathcal{J}A$. This completes the proof when p has an associate that is not in $\boldsymbol{D}$.

Case 2: Suppose $q \in \boldsymbol{D}$. From the work in §6.2 and §12.5 we have the following situation.

- The Cantor set $\mathcal{J}C$ is the nested intersection of the sets $\boldsymbol{K}(n)$. Here $\boldsymbol{K}(n)$ consists of 6^n disjoint disks, each containing 6 disks of $\boldsymbol{K}(n+1)$.
- $\boldsymbol{K}(0) = \boldsymbol{D}$ is the big disk we discussed at length in §12.2.
- $\boldsymbol{K}(1) \subset \boldsymbol{D}_1 \cup \cdots \cup \boldsymbol{D}_6$, the 6 disks discussed in §12.3.
- $\boldsymbol{H}$ maps $\boldsymbol{K}(m+1)$ to $\boldsymbol{K}(m)$ and $\boldsymbol{H}$ maps each point of $\boldsymbol{D} - \boldsymbol{K}(1)$ into $\boldsymbol{R}^2 - \boldsymbol{D}$ provided $\boldsymbol{H}$ is defined at this point.
- The point $\mathbf{s}$ is the special point defined in connection with Figure 12.4, and mentioned in Lemma 12.4.

Replacing p by a suitable image of p under the Gauss group, we can assume that $q = \boldsymbol{B}(p)$. Since $p \notin \mathcal{J}C$, there is some $m \geq 0$ such that $q \in \boldsymbol{K}(m) - \boldsymbol{K}(m+1)$. Since all the primary elements are defined and finite on the sets $\boldsymbol{K}(1), \boldsymbol{K}(2), \ldots$, we see that $\boldsymbol{H}^m(q) \in \boldsymbol{D} - \boldsymbol{K}(1)$. Replacing p by $H^m(p)$, we can assume that

$$q = \boldsymbol{B}(p) \in \boldsymbol{D} - \boldsymbol{K}(1). \tag{14.9}$$

There are now two sub-cases.

Case 2A: Suppose $q \in \bigcup \boldsymbol{D}_k - \boldsymbol{K}(1)$. Since all primary maps are defined and finite on $\bigcup \boldsymbol{D}_k$ we have

$$\boldsymbol{H}(q) \in \boldsymbol{R}^2 - \boldsymbol{D}. \tag{14.10}$$

Since $\boldsymbol{D}$ is $\boldsymbol{\Gamma}$-invariant, we see that $H(p)$ cannot have an associate in $\boldsymbol{D}$. Replacing p with $H(p)$ we reduce to Case 1.

Case 2B: If $q \notin \bigcup \boldsymbol{D}_k$. Then $q \in \boldsymbol{D} - \bigcup \boldsymbol{D}_k$. We have scramble here because $\boldsymbol{H}$ might blow up at q. We aim to apply Lemma 12.4. Since p has a well-defined H-orbit and $q = \boldsymbol{B}(p)$, and $\boldsymbol{B}$ conjugates H to $\boldsymbol{H}$, the condition $q \in \boldsymbol{\Gamma}(\mathbf{s})$ would force $p \in \Gamma(s)$, where $s = \boldsymbol{B}^{-1}(\mathbf{s})$. But points in $\Gamma(s)$ do not have well-defined orbits. In short, $q \notin \boldsymbol{\Gamma}(\mathbf{s})$.

By Lemma 12.4, there is some primary element h, well defined and finite at q, such that

$$h(q) \in \boldsymbol{R}^2 - \boldsymbol{D}. \tag{14.11}$$

Any associate of $H(p)$ is in the same $\boldsymbol{\Gamma}$ orbit as $h(q)$. But, since $\boldsymbol{D}$ is $\boldsymbol{\Gamma}$-invariant, this means that $H(p)$ cannot have an associate in $\boldsymbol{D}$. Replacing p by $H(p)$ we again reduce to Case 1.

This completes the proof of equation (14.7). Our proof of Theorem 1.6 is done.

Part 3

CHAPTER 15

Sketches for the Remaining Results

In this chapter we will provide detailed sketches for the structural results in §1.5. Some readers would probably appreciate getting an account of the ideas behind the proofs without having to wade through the details. Overall the ideas are pretty simple, once the complicating details are stripped off. We persist in the practice of prepending a $\heartsuit$ to any subset of the Julia set in $\boldsymbol{B}$-coordinates.

15.1. The General Setup

Before we get started with the sketches, we recall the general setup.

Our basic map is $\boldsymbol{H} = \boldsymbol{B}H\boldsymbol{B}^{-1}$ where H is the projective heat map from equation (1.1) and $\boldsymbol{B}$ is the change of coordinates map from equation (1.2). We consider the action of $\boldsymbol{H}$ on the manifold $\boldsymbol{M}$ obtained by blowing up $(\boldsymbol{R} \cup \infty)^2$ at the three points

$$(\phi^2, -\phi^4), \qquad (-\phi^4, \phi^2), \qquad (\phi^6, \phi^6).$$

The Gauss group $\boldsymbol{\Gamma}$ acts as a group of diffeomorphisms of $\boldsymbol{M}$. Here we have $\boldsymbol{\Gamma} = \boldsymbol{B}\Gamma\boldsymbol{B}^{-1}$, where Γ is the order 10 group discussed in §3.8. The group $\boldsymbol{\Gamma}$ is generated by $\boldsymbol{R}(x,y) = (y,x)$ and $\boldsymbol{G} = \boldsymbol{B}G\boldsymbol{B}^{-1}$, where G is the Gauss recurrence defined in equation (3.16). We recall the notation

$$\boldsymbol{H}_{ijk} = \boldsymbol{R}^i\boldsymbol{G}^i\boldsymbol{H}^k \tag{15.1}$$

The elements $\boldsymbol{H}_{ij1}$ are called the *primary elements*.

Let $\boldsymbol{T}$ be the fundamental domain from Part 2, the triangle with vertices

$$(0,0), \qquad (\phi^6, -\phi^4), \qquad (\phi^6, \phi^6).$$

A fundamental domain for the action of $\boldsymbol{\Gamma}$ on $\boldsymbol{M}$ is obtained by blowing up $\boldsymbol{T}$ at the vertex (ϕ^6, ϕ^6). (The other blow-up points do not lie in $\boldsymbol{T}$.) The result is a quadrilateral, in which 3 of the sides are Euclidean line segments and the fourth side is a segment of the exceptional fiber lying over the (ϕ^6, ϕ^6). We call this fundamental domain $\boldsymbol{T}$ as well. See Figures 10.4 and 10.5.

We let $\heartsuit J$ denote the closure of the set of points in $\boldsymbol{M}$ which have well-defined $\boldsymbol{H}$ orbits and which do not converge to (∞, ∞), the point representing the regular class. The set $\heartsuit J$ is the *Julia set*. It is the main object of study in this part of the monograph.

First we recall some important subsets we have already identified. $\boldsymbol{D}$ is the big disk used in the proofs of Theorems 1.5 and 1.6. So far, we have identified a Cantor set $\heartsuit JC \subset \heartsuit J \cap \boldsymbol{D}$ and the Cantor band $\heartsuit JA \subset \heartsuit J - \boldsymbol{D}$. Here $\heartsuit JC$ and $\heartsuit JA$ are the images under $\boldsymbol{B}$ of the sets $\mathcal{J}C$ and $\mathcal{J}A$ from Theorems 1.5 and 1.6 respectively. We also have the smaller Cantor band $\heartsuit A = \heartsuit JA \cap \boldsymbol{T}$. The Cantor band $\heartsuit JA$ is the $\boldsymbol{\Gamma}$ orbit of $\heartsuit A$.

15.2. The Solenoid Result

Here we sketch the proof of Theorem 1.10, which says that $\heartsuit J$ contains a set which, when blown up at all its cone points, is homeomorphic to the 5-fold cover of the 2-adic solenoid. The set in Theorem 1.10 is $\heartsuit J - \boldsymbol{D}$ together with the 5 vertices of $\boldsymbol{D}$. The details are done in §16 and §17.

Let $F : \clubsuit P \to \boldsymbol{R}^2$ be the quasi-horseshoe from Theorem 14.1. The domain $\clubsuit P$ of F is the union of 8 quadrilaterals shown in Figure 13.1. The Cantor band $\heartsuit A$ consists of those points $x \in \clubsuit P$ such that $F^m(x) \in \clubsuit P$ for all $n = 1, 2, 3, \ldots$.

The domain $\clubsuit P$ is flanked on either side by some of the other pieces of the partition, namely the pieces $\clubsuit I, \clubsuit J, \clubsuit K, \clubsuit L$. These pieces are further subdivided – e.g. $\clubsuit L = \clubsuit L_1 \cup \clubsuit L_2$. Let $\clubsuit P^*$ denote the union of all these pieces, including $\clubsuit P$. Thus $\clubsuit P \subset \clubsuit P^*$. We have associated primary maps to each of the pieces in $\clubsuit P$, and we now define $F : \clubsuit P^* \to \boldsymbol{R}^2$ simply by letting the associated primary element act on each piece. The set $\clubsuit P^*$ comes in 4 rows, and the same primary element acts in each row.

We define

$$\heartsuit A^* = F^{-1}(\heartsuit A) \cup F^{-2}(\heartsuit A). \tag{15.2}$$

It turns out that $\heartsuit A^*$ is precisely the set of points $x \in \clubsuit P^*$ so that $F^n(x) \in \clubsuit P$ for all $n > 2$. The remaining points in $\clubsuit P^*$ are preconvex. In other words

$$\heartsuit A^* = \heartsuit J \cap \clubsuit P^*. \tag{15.3}$$

We analyze $\heartsuit A^*$ by looking at the properties of the differential $d\boldsymbol{F}$. In the next section we will go in to more detail on this point.

It turns out that $\heartsuit A^*$ is obtained from $\heartsuit A$ just by extending the strands of $\heartsuit A$ until they hit the boundary of $\boldsymbol{T}$ on either side. Figure 15.1 shows a schematic picture. $\clubsuit P$ is the shaded region. $\clubsuit P$ naturally comes in 4 layers. The pieces of $\clubsuit P^*$ are not drawn.

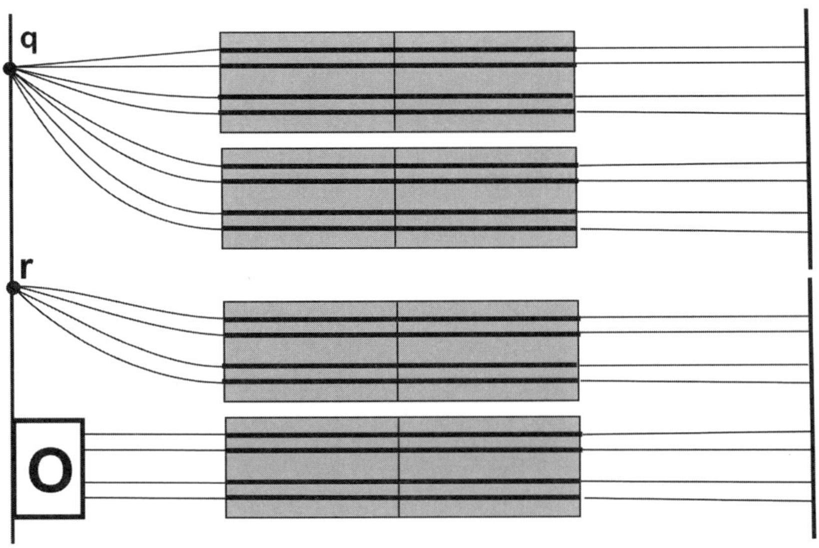

FIGURE 15.1. $\heartsuit A$ and $\heartsuit A^*$

We establish the following structure:

(1) The strands of $\heartsuit A^*$ hit the right edge $\partial_1 \boldsymbol{T}$ of $\boldsymbol{T}$ transversely in a Cantor set $\heartsuit A_R^*$.

(2) The top half of $\heartsuit A^*$ pinches down to a point $\boldsymbol{q}$ on the left boundary $\partial_0 \boldsymbol{T}$ of $\boldsymbol{T}$.

(3) The top half of the bottom half of $\heartsuit A^*$ pinches down to a point $\boldsymbol{r}$ on the left boundary of $\boldsymbol{T}$.

(4) The bottom half of the bottom half of $\heartsuit A^*$ intersects the partition piece $\clubsuit O$ in a Cantor set $\heartsuit A_L^*$.

Let $h = \boldsymbol{H}_{101}$. We already know that $h : \clubsuit O \to \boldsymbol{T}$ is expanding. We check that $h(\heartsuit A_L^*) = \heartsuit A_R^*$. We set $\heartsuit A_k^* = h^{-k}(\heartsuit A_0^*)$ for $k = 0, 1, 2, \ldots$. The right ends of $\heartsuit A_{k+1}^*$ match the left ends of the bottom fourth of $\heartsuit A_k^*$ for all k. Figure 15.2 shows a schematic picture of how these pieces fit together. In Figure 15.2, the biggest dark piece is $\heartsuit A_0^*$. The top big dark triangle represents the top half of $\heartsuit A_0^*$. The second big dark triangle represents the top half of the bottom half of $\heartsuit A_0^*$. The big dark rectangle represents the bottom half of the bottom half of $\heartsuit A_0^*$. The other pieces are interpreted similarly.

A *Cantor cone* is the cone on a Cantor set. We will show that $\bigcup \heartsuit A_k^*$ is an infinite union of Cantor cones having their cone points on the left edge of $\boldsymbol{T}$. The

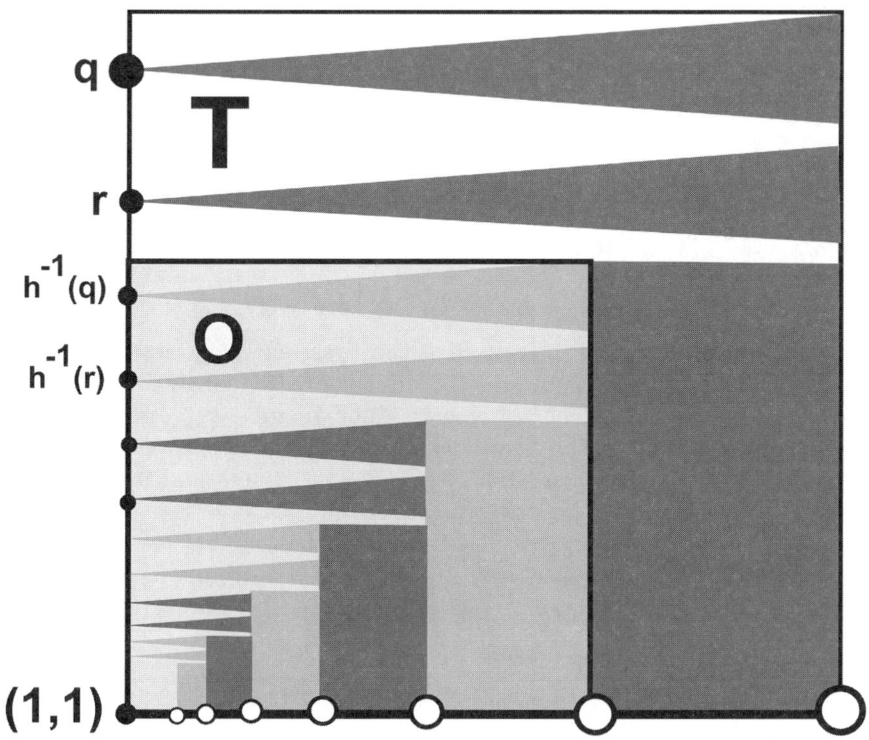

FIGURE 15.2. $\heartsuit A_0^* \cup \cdots \cup \heartsuit A_6^*$ shown schematically

set $\heartsuit J \cap \boldsymbol{T} - \clubsuit Q$ is this union, together with one half-open arc that goes along the bottom. The endpoint $(1,1)$ of this arc is a vertex of $\boldsymbol{D}$. When we blow up all the cone points and adjoin $(1,1)$, we get a product $S \times [0,1]$ where S is a Cantor set and p is a single deleted point. The action of $\boldsymbol{\Gamma}$ induces some identifications on the ends of $S \times [0,1]$. When we check the identifications we find that the quotient is the BJK continuum. When we take the 10-fold cover we get the 5-fold cover of the 2-adic solenoid.

15.3. Local Structure

Here we sketch the proof of Theorem 1.8, which says that $\heartsuit J$ is the union of a Cantor set (namely $\heartsuit JC$), a countable collection of Cantor bands, and a countable collection of cone points. The details are done in §18.

There is one notational remark we make before getting started. The disk $\boldsymbol{D}$ is a subset of $\boldsymbol{M}$ but none of the exceptional fibers of $\boldsymbol{M}$ intersects $\boldsymbol{D}$. For this reason the blow-down map $\pi : \boldsymbol{M} \to (\boldsymbol{R} \cup \infty)^2$ is a diffeomorphism on $\boldsymbol{D}$. For this reason we will often confuse $\boldsymbol{D}$ and $\pi(\boldsymbol{D})$. They are essentially the same set.

Call a point of $\heartsuit J$ a *band point* if it has a neighborhood which intersects $\heartsuit J$ in a Cantor band. Our work in the proof of Theorem 1.10 shows that $\heartsuit J - \boldsymbol{D}$ consists entirely of cone points and band points. We just have to show that $\heartsuit J \cap \boldsymbol{D} - \heartsuit JC$ consists entirely of cone points and band points.

As a first step, we will show that the set

$$\pi(\heartsuit J) \cap \boldsymbol{R}^2 - \boldsymbol{D} \tag{15.4}$$

consists of cone points and band points. The set in Equation 15.4 is almost the same set as the set $\heartsuit J \cap \pi^{-1}(\boldsymbol{R}^2) - \boldsymbol{D}$. The only difference is that we have to make sure that nothing strange happens when we blow down the exceptional fibers.

Once we know that the set in Equation 15.4 consists entirely of cone points and band points, we will use the maps and partition pieces that we used in the proof of Theorem 1.5. We will have three general situations. The first situation is that there is some partition piece $\clubsuit X$ and some primary map $h : \clubsuit X \to \boldsymbol{R}^2 - \boldsymbol{D}$ which is a local diffeomorphism. In this case, we can conclude that $\heartsuit J \cap \clubsuit X$ consists entirely of cone points and band points, because

$$h(\heartsuit J \cap \clubsuit X) \subset \pi(\heartsuit J) \cap \boldsymbol{R}^2 - \boldsymbol{D}.$$

The second situation we have is like the first, except that h is not a local diffeomorphism. In these cases, we will find a foliation of $\clubsuit X$ by straight line segments such that h maps each segment to a non-singular curve that is transverse to any of the arcs in $\heartsuit J$ contained in $h(\clubsuit X)$. In this situation we will also be able to conclude that $\heartsuit J \cap \clubsuit X$ consists entirely of cone points and band points. We will analyze this situation using the cone test from §11.

We can almost get away with just the first situation, but we run out of luck when we deal with the piece $\clubsuit S_1$ from the proof of Theorem 1.5. We will have to divide $\clubsuit S_1$ into 7 pieces and treat each one separately. The two out of 7 which require the second situation are painful to deal with.

The methods above take care of every point of $\heartsuit J - \heartsuit JC$ except for 5 special points in a single $\boldsymbol{\Gamma}$ orbit. By symmetry, we just have to analyze the local structure around one more point. We do this at the end. Again, it is rather painful. In short, we prove Theorem 1.8 by pulling back the information we get from Theorem 1.10 in an inductive way.

Our methods establish the stronger fact that all the strands of $\heartsuit J$ are $\boldsymbol{C}^1$-arcs and many of the cone points are what we call *basic*. By this we mean that the tangent vectors of the strands make sense even at the cone point, and one can continue the strands through the cone point uniquely in a C^1 way. In other words, when we blow up at the cone point, the subset of $\heartsuit J$ near the point becomes a C^1 Cantor band. These basic cone points come from pulling back the ones in $\heartsuit J - \boldsymbol{D}$. This C^1 structure is a key ingredient in the proofs of the remaining results.

15.4. The Embedded Graph

Here we sketch the proof of Theorem 1.9, which says that $\heartsuit J$ contains a homeomorphic copy of the infinite graph $\mathcal{G}$ suggested by Figure 1.5. The details are done in §19.

Our construction refers to Figure 15.3, which shows a schematic picture. The outer grey pentagon represents the disk $\boldsymbol{D}$. The black vertices are the 5 fixed points of $\boldsymbol{H}$ which are neither the regular nor the star-regular class. (We call these the *irregular* fixed points.) The 6 inner grey pentagons represent the 6 disks $\boldsymbol{D}_1 \cup \ldots \boldsymbol{D}_6$. The black pentagon represents a set $\boldsymbol{\Gamma}(\boldsymbol{Y})$, which we describe below. The shaded triangle represents the portion contained in $\boldsymbol{T}$.

Our argument has 3 main parts. In the first part we isolate a single embedded arc $\boldsymbol{X} \subset \heartsuit J$. This is the arc connecting points a_1 and a_2. Figure 15.3. We call $\boldsymbol{X}$ *the generator*.

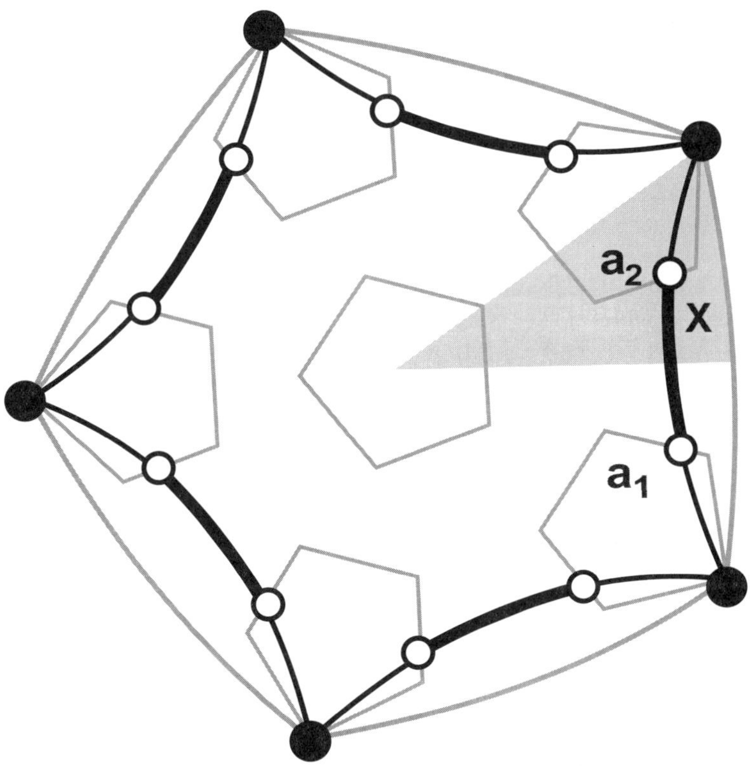

FIGURE 15.3. The arc $\boldsymbol{X}$ and the loop $\boldsymbol{\Gamma}(\boldsymbol{Y})$

In the second part, we look at the closure of a certain infinite union of pre-images of $\boldsymbol{X}$ under elements of our semigroup. We call this set $\boldsymbol{Y}$. We show that $\boldsymbol{Y}$ is an embedded arc in $\heartsuit J$ which connects the two irregular fixed points of $\boldsymbol{H}$, and that the orbit $\boldsymbol{\Gamma}(\boldsymbol{Y})$ is an embedded loop. In §20 we will show that this loop is canonical, in the sense that it bounds that smallest disk that contains $\heartsuit J \cap \boldsymbol{D}$.

In the third part, we consider the 6 pre-images of $\boldsymbol{\Gamma}(\boldsymbol{Y})$. The region bounded by $\boldsymbol{\Gamma}(\boldsymbol{Y})$ and these 6 smaller loops is topologically a disk with 6 smaller disks removed. We find a collection of 30 arcs running through this topological disk in a pattern that matches the left side of Figure 1.5. We really just find a certain collection $\boldsymbol{Z}$ of 3 arcs. The remaining arcs are in the orbit $\boldsymbol{\Gamma}(\boldsymbol{Z})$. We recognize $\boldsymbol{\Gamma}(\boldsymbol{Y} \cup \boldsymbol{Z})$ as a copy of the seed for $\mathcal{G}$ shown on the left side of Figure 1.5. We get $\mathcal{G}$ itself by taking all the pre-images under $\boldsymbol{H}$.

15.5. Path Connectivity

Here is a sketch of our proof of Theorem 1.11, which shows that $\heartsuit J$ is path connected. The details are done in §20.

We already know that the closure of $\heartsuit J - \boldsymbol{D}$ is path connected. Let $\heartsuit G$ denote the homeomorphic copy of the graph $\mathcal{G}$ from Theorem 1.9. The set $\heartsuit G$ is also path connected, and it meets the closure of $\heartsuit J - \boldsymbol{D}$ at the 5 irregular fixed points of $\boldsymbol{H}$. So, to finish the proof, we just have to show that every point of $\heartsuit J \cap \boldsymbol{D}$ can be connected to a point of $\heartsuit G$ by a path.

As a first step, in §20 we show that the disk $\boldsymbol{L}$ bounded by $\boldsymbol{\Gamma}(\boldsymbol{Y})$ contains $\heartsuit J \cap \boldsymbol{D}$. (We already mentioned this above.) This result is nice to know on its own, and also it helps eliminate some of the places where additional connected components of $\heartsuit J$ might be hiding. After we know that $\heartsuit J \cap \boldsymbol{D} \subset \boldsymbol{L}$, we know that any additional connected component must lie in $\boldsymbol{L}$.

We will see that we just have to consider points inside $\boldsymbol{L}$ and outside the pre-images of $\boldsymbol{H}^{-1}(\boldsymbol{L})$. Every point of $\heartsuit J$ in this set is either a cone point or a band point. In either case we just follow along a strand of $\heartsuit J$ which contains the point and we prove that the path we take must end of $\heartsuit G$.

Here we explain the main idea of the approach. Suppose we have some path component Ψ which is not connected to $\heartsuit JC$. We choose some C^1 band point in Ψ and continue the strand containing it in both directions. We will show that we can always continue the strand through cone points. This gives us an extended strand β which is either a C^1 loop or a C^1 bi-infinite path. At the same time, we will always have a primary element $h : \beta \to \boldsymbol{R}^2$ such that $h(\beta) \subset \heartsuit J - \boldsymbol{D}$. We will show that $\heartsuit J - \boldsymbol{D}$ has no C^1 loops and also that no bi-infinite path in $\heartsuit J - \boldsymbol{D}$ can be a subset of $\boldsymbol{R}^2$. This gives a contradiction: $h(\beta)$ cannot exist.

15.6. The Postcritical Set

Here we sketch a proof that the map H is not post-critically finite. We show this for $\boldsymbol{H}$ instead, which is equivalent. We work in the manifold $\boldsymbol{M}$. Unlike the sketches in previous chapters, these are all the details we will supply.

The *critical set* is the set $\boldsymbol{Y} \subset \boldsymbol{R}^2$ where $d\boldsymbol{H}$ is singular. Since the elements of $\boldsymbol{\Gamma}$ act on $\boldsymbol{M}$ by diffeomorphisms, the critical set in $\boldsymbol{M}$ is the same for every primary map, including $\boldsymbol{H}$. We look carefully at the partition piece $\clubsuit K_2$ which lies to the left of the set $\boldsymbol{P}$ containing the quasi-horseshoe.

LEMMA 15.1. $\boldsymbol{Y}$ *contains a connected smooth arc which intersects more than one strand of* $\heartsuit J \cap \clubsuit K_2$.

Proof: We work with the primary element $h = \boldsymbol{H}_{041}$. This map is defined on all of $\clubsuit K_2$ and maps it into $\boldsymbol{R}^2$ avoiding the blow-up points. So, we can decide whether $d\boldsymbol{H}$ is singular just by treating h as a smooth map from a subset of $\boldsymbol{R}^2$ into $\boldsymbol{R}^2$. To prove what we want, we find 2 vertical line segments $\nu_+, \nu_- \in \clubsuit K_2$ such that

- $\det(dh) < 0$ on ν_- and $\det(dh) > 0$ on ν_+.
- One endpoint of $\nu_\pm$ lies above some of the strands of $\heartsuit A$ and the other endpoint lies below the same strands.

One can take ν_+ to be the right edge of $\clubsuit K_2$ and ν_- to be the bottom 3/4 of the left edge of $\clubsuit K_2$. □

Our smooth arc of $\boldsymbol{Y}$ cannot be everywhere tangent to $\heartsuit J$ because it intersects more than one strand. So $\boldsymbol{Y}$ contains a smaller smooth arc which intersects a strand of $\heartsuit J$ transversely.

The forward image of $\boldsymbol{Y}$ under a suitable composition of primary elements, specifically $\boldsymbol{H}_{111}^2$, is a continuous curve which intersects the Cantor band $\heartsuit A$ in infinitely many arcs. The argument used in Lemma 6.7 shows that, for any n, some forward image of $\boldsymbol{Y}$ contains union of n smooth arcs $\boldsymbol{Y}_1, \ldots, \boldsymbol{Y}_n$ with the following properties.

(1) Each $\boldsymbol{Y}_j$ stretches entirely across $\heartsuit A$.
(2) Each $\boldsymbol{Y}_j$ is timelike.
(3) For some strand σ, not the top or the bottom strand, the intersection points $\boldsymbol{Y}_1 \cap \sigma, \ldots, \boldsymbol{Y}_n \cap \sigma$ are all distinct and within $1/n$ of each other.

In this case, we can find a continuous family of parallel lines (all nearly parallel to σ and near the intersection points) which each intersect the forward image of $\boldsymbol{Y}$ in n distinct points. But then, by Bezout's Theorem, there is no upper bound on the degree of $\boldsymbol{Y}$. This makes it impossible for $\boldsymbol{Y}$ to lie in a finite union of algebraic curves.

15.7. No Rational Fibration

Here we sketch a proof that there is no nontrivial pair (f, h) where $f : \boldsymbol{R}^2 \to \boldsymbol{R}$ and $h : \boldsymbol{R} \to \boldsymbol{R}$ are rational and $fH = hf$. Again, these are all the details on this we are going to provide.

We will argue by contradiction. Suppose that $f : \boldsymbol{R}^2 \to \boldsymbol{R}$ is such a map. The main property we will use is that H maps fibers of f into fibers of f. preserves the fibers of f.

$\heartsuit J - \boldsymbol{D}$ has a decomposition into C^1 arcs and cone points. Suppose first that f is constant along the C^1 arcs of $\heartsuit J - \boldsymbol{D}$.

LEMMA 15.2. f *takes on the same value on any two* C^1 *arcs of* $\heartsuit J$ *which meet at a cone point and have the same tangent line at the cone point.*

Proof: Consider the behavior of f at a cone point (x_0, y_0). In local coordinates we have

$$f(x,y) = \frac{P_0(x-x_0, y-y_0) + P_1(x-x_0, y-y_0)}{Q_0(x-x_0, y-y_0) + Q_1(x-x_0, y-y_0)} \tag{15.5}$$

Where P_0 and Q_0 are homogeneous polynomials obtained by collecting the lowest order terms. P_1 and Q_1 are the higher order terms. Since f is constant along the C^1 arcs of $\heartsuit J - \boldsymbol{D}$, we see that f has uncountably many well defined directional limits as $(x, y) \to (x_0, y_0)$. Not all the directional limits can be 0 because then f would be identically 0. This situation forces P_0 and Q_0 have the same degree. So, if (x', y') is chosen so that (x_0, y_0) is the midpoint of the segment joining (x, y) to (x', y'), we have

$$\frac{P_0(x' - x_0, y' - y_0)}{Q_0(x' - x_0, y' - y_0)} = \frac{P_0(x - x_0, y - y_0)}{Q_0(x - x_0, y - y_0)}. \tag{15.6}$$

So, $f(x, y) = f(x', y')$ up to terms which vanish when these points converge to the cone point. □

The previous result tells us that f is constant on one of the infinite C^1 paths mentioned above. These paths are dense in $\heartsuit J - \boldsymbol{D}$, and this forces f to be constant on $\heartsuit J - \boldsymbol{D}$. But then f is constant.

So, f is not always constant along the arcs of $\heartsuit J - \boldsymbol{D}$. But then there is some level curve which, at least in some small open set, is smooth and transverse to $\heartsuit J$ at some point of $\heartsuit J - \boldsymbol{D}$. Pushing forward by $\boldsymbol{H}$ finitely many times, we can find a level curve λ which is transverse to $\heartsuit JA$. Suppose for convenience that λ is transverse to $\heartsuit A$. (It is more convenient to work with $\heartsuit A$ rather than one of the other $\boldsymbol{\Gamma}$ translates which comprise $\heartsuit JA$, but not essential to the argument.)

Now we argue as in §15.6. When we push forward our transverse curve by a high power of $\boldsymbol{H}$, it intersects the same strand of $\heartsuit A$ many times, transversely. This makes the image of the curve have arbitrarily high degree. On the other hand, the push-forward of the curve supposedly is just some other level curve for the fibration. This is a contradiction.

CHAPTER 16

Towards the Solenoid

16.1. The Four Strips

The goal of the next 2 chapters is to prove Theorem 1.10. We follow the sketch given in §15.2. In this chapter we produce the set $\heartsuit A^*$ from §15.2.

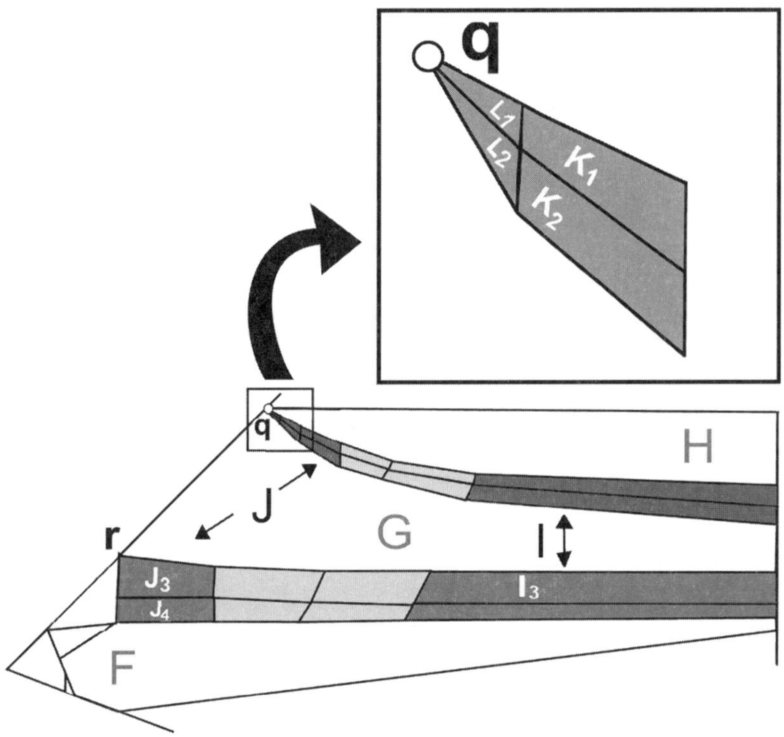

FIGURE 16.1. Figure 13.5 repeated

Figure 16.1 shows a repeat of Figure 13.5. This figure is central to our analysis. We organize the pieces shown in Figure 16.1 into 4 rows, as follows.

(1) $\clubsuit P_k^* = \clubsuit L_k \cup \clubsuit K_k \cup \clubsuit J_1 \cup \clubsuit P_k \cup \clubsuit I_k$ for $k = 1, 2$.

(2) $\clubsuit P_k^* = \clubsuit J_k \cup \clubsuit P_k \cup \clubsuit I_k$ for $k = 3, 4$.

We call these 4 sets *strips*. We omit the points $\boldsymbol{q}$ and $\boldsymbol{r}$ from $\clubsuit P^*$ because they play a special role in the constructions.

Define

$$\clubsuit P^* = \bigcup_{i=1}^{4} \clubsuit P_i^*. \tag{16.1}$$

$\clubsuit P^*$ is precisely the "Inner Layer" of pieces we considered in §13.3. As in §14, we define

$$h_1 = \boldsymbol{H}_{111}, \quad h_2 = \boldsymbol{H}_{041}, \quad h_3 = \boldsymbol{H}_{141}, \quad h_4 = \boldsymbol{H}_{021}. \tag{16.2}$$

and we let $F : \clubsuit P^* \to \boldsymbol{R}^2$ be such that the restriction of F to $\clubsuit P_i^*$ is h_i. As in §14, we adjust the vertices by a tiny constant so that the 4 strips are pairwise disjoint. (We don't move $\boldsymbol{q}$ or $\boldsymbol{r}$ however.) This avoids conflicts in the definitions on the common boundaries.

Letting $\heartsuit A$ be the Cantor band corresponding to the quasi-horseshoe from Theorem 14.1, we define

$$\heartsuit A^* = \Big(F^{-1}(\heartsuit A) \cup F^{-2}(\heartsuit A)\Big) \cap \clubsuit P^*. \tag{16.3}$$

First we study the structure of $\heartsuit A^*$ and then we pull $\heartsuit A^*$ back using inverse powers of the primary element $h = \boldsymbol{H}_{101}$. When we are done, we will have have our countable collection of Cantor cones.

16.2. Two Cantor Cones

16.2.1. The Right Half. We first study the restriction of F to $\clubsuit I$. Say that a *Cantor product* is a space homeomorphic to the product of a closed interval and a Cantor set. A Cantor band is what you get when you remove all the endpoints of the strands from a Cantor product.

LEMMA 16.1. *$\heartsuit A^* \cap \clubsuit I$ is a Cantor product whose strands connect one vertical side of $\clubsuit I$ to the other. All the stands in $\heartsuit A^* \cap (\clubsuit I \cup \clubsuit P)$ have slope in the interval* $[-1/2, 1/2]$.

Proof: We now make 4 observations.

1. We use the area test to check that F is a local diffeomorphism on $\clubsuit I$.
2. From the work in §14, the set $F(\clubsuit I)$ is contained in a quadrilateral $\spadesuit Y_1$ which interlaces $\clubsuit P$. See Figure 14.1.
3. From the work in §14, the set $F(\clubsuit I) - \clubsuit P$ consists of pre-convex points.
4. We use the exclusion test to check that F maps the top and bottom edges of each piece of $\clubsuit I$ outside the quadrilateral $\spadesuit Z$. See Figure 14.1. Since $\clubsuit P^* \subset \spadesuit Z$ we see that F stretches each piece of $\clubsuit I$ completely over $\clubsuit P^*$.

Figure 16.2 shows a schematic picture for the top piece $\clubsuit I_1$. The picture is similar schematically for the other 3 pieces.

From Item 1, we conclude that $F^{-1}(\heartsuit A) \cap I'$ is locally a Cantor band. Here I' is some small open set which contains $\clubsuit I$. (We consider I' so we don't have to talk specially about the vertical sides of $\clubsuit I$.) From Item 2, we conclude that each strand of $F^{-1}(\heartsuit A) \cap \clubsuit I$ can only exit $\clubsuit I$ through a vertical side of $\clubsuit I$. From Item 3 and the forward invariance of $\heartsuit A$, we conclude that $F^{-2}(\heartsuit A) \subset F^{-1}(\heartsuit A)$. Hence $F^{-1}(\heartsuit A) \cap \clubsuit I = \heartsuit A^* \cap I$. From Item 4 we conclude that no strand of $\heartsuit A^* \cap \clubsuit I$ intersects the top or bottom of a piece of $\clubsuit I$. From all this we see that

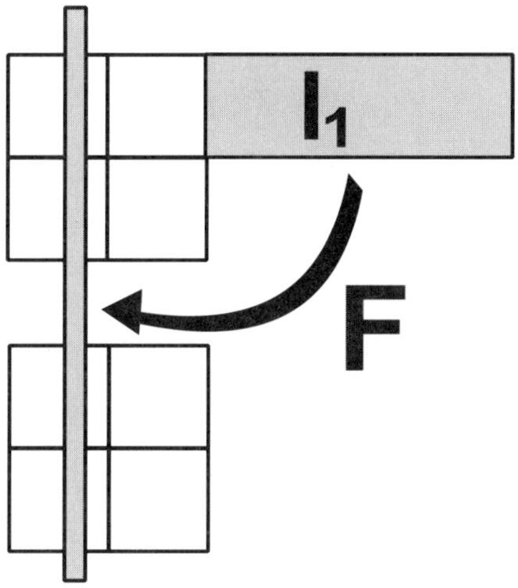

FIGURE 16.2. A schematic picture of $\clubsuit I_1$ and $F(\clubsuit I_1)$ and $\clubsuit P$

$\heartsuit A^* \cap I'$ is a Cantor band. Hence $\heartsuit A^* \cap \clubsuit I$ is a Cantor product which has its endpoints on the vertical sides of $\clubsuit I$.

For the claim about the slopes, we use the cone test to show that

$$dF(\vee(1/2,-1/2)) \subset \vee \tag{16.4}$$

throughout $\clubsuit I \cup \clubsuit P$. Since the strands of $\heartsuit A$ are spacelike with respect to $\vee$, this proves what we want. □

16.2.2. The Bottom Left. Now we see what F does to $\clubsuit J_3$ and $\clubsuit J_4$. Note that $\clubsuit J_3$ has $\boldsymbol{r}$ as a vertex, and this point requires special treatment.

LEMMA 16.2. *$\heartsuit A^* \cap \clubsuit J_4$ is a Cantor product whose strands connect the vertical sides of $\clubsuit J_4$.*

Proof: The proof is just like we gave above, except that in place of Item 2 we use the statement that $F(\clubsuit J_4)$ is contained in a quadrilateral which interlaces the polygon $\clubsuit P \cup \clubsuit I$, as shown schematically in Figure 16.3. □

LEMMA 16.3. *The closure of $\heartsuit A^* \cap (\clubsuit J_3 - \boldsymbol{r})$ is a Cantor cone whose strands connect the right vertical side of $\clubsuit J_3$ to the cone point* $\mathbf{r}$

Proof: The proof is again the same except that we need to be clear what we mean by the top and bottom sides. We declare the top side of $\clubsuit J_3$ to be the single side which connects $\boldsymbol{r}$ to the vertical side of $\clubsuit P$. We declare the bottom "side" of $\clubsuit J_3$ to be the union of the other 2 sides which connect $\boldsymbol{r}$ to the vertical side of $\clubsuit P$. Figure 16.4 shows what we mean. With this definition in place, we check in the same way that F stretches $\clubsuit J_3$ over $\clubsuit P^*$ in the same sense as F stretches F_4 over $\clubsuit P$, except that the map is not defined at $\mathbf{r}$. The same arguments as above now finish the proof. The strands must limit on $\boldsymbol{r}$ because they cannot exit the top or bottom sides, and these sides both limit on $\boldsymbol{r}$. □

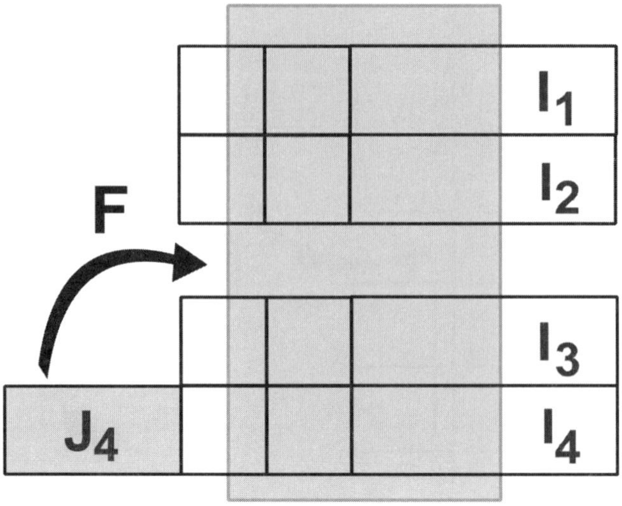

FIGURE 16.3. A schematic picture of $\clubsuit J_4$ and $F(\clubsuit J_4)$ and $\clubsuit P$

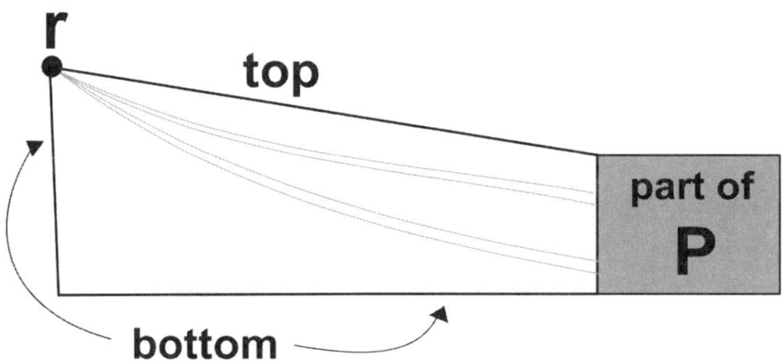

FIGURE 16.4. The piece $\clubsuit J_3$

16.2.3. The Top Left. Finally, we see what F does to the pieces

$$\clubsuit L, \clubsuit K, \clubsuit J_1, \clubsuit J_2$$

in Figure 16.1. Our luck runs out because dF is sometimes singular in these pieces. See the discussion in §15.6. We will use a different trick.

LEMMA 16.4. *Let $\clubsuit X$ be one of $\clubsuit J_1, \clubsuit J_2, \clubsuit K_1, \clubsuit K_2$. Then $\heartsuit A^* \cap \clubsuit X$ is a Cantor product whose strands connect the vertical sides of $\clubsuit X$.*

Proof: We check that F stretches $\clubsuit X$ over $\clubsuit P$ and that $F(\clubsuit X)$ interlaces one of the two columns of squares in $\clubsuit P$. Since $\clubsuit X$ has two opposite vertical edges, we can consider the foliation of $\clubsuit X$ by vertical line segments. Let σ be such a segment. We use the cone test to check for each $p \in \clubsuit X$ that $dF_p(0,1)$ never has slope in $(-1,1)$. That is, $dF_p(0,1)$ is not spacelike. Since $\heartsuit A$ consists of spacelike curves, we see that $F(\sigma)$ is a smooth nonsingular arc which passes through the top and bottom of $\clubsuit P$ and is always transverse to $\heartsuit A$. Hence $F^{-1}(\heartsuit A) \cap \clubsuit X$ is a Cantor set. Since the strands of $\heartsuit A$ are continuously differentiable and $dF_p(0,1)$ varies smoothly we see by transversality that the intersection $\sigma \cap \heartsuit A^*$ varies continuously with σ. This shows that $\clubsuit X \cap \heartsuit A^*$ is a Cantor product whose strands connect one vertical side of $\clubsuit X$ to the other. $\square$

LEMMA 16.5. *Let $\clubsuit X$ be either $\clubsuit L_1$ or $\clubsuit L_2$. The closure of $\heartsuit A^* \cap (\clubsuit X - \boldsymbol{q})$ is a Cantor cone whose strands connect the vertical side of $\clubsuit X$ to q.*

Proof: This has the same proof as for the previous result. This time $\clubsuit X$ is a triangle with a vertical side, a top side, and a bottom side. We use the weak cone test this time. $\square$

Our next result is essentially an application of transversality.

LEMMA 16.6. *The strands of $\heartsuit A^* - \boldsymbol{q} - \boldsymbol{r}$ are C^1. That is, they are continuously differentiable.*

Proof: The strands of $\heartsuit A^* \cap \clubsuit I$ are C^1 because the strands of $\heartsuit A$ are C^1 and F is a local diffeomorphism on $\clubsuit I$. Hence the strands of $\heartsuit A^* \cap (\clubsuit P \cup \clubsuit I)$ are C^1. This remains true even when they cross the interface between $\clubsuit I$ and $\clubsuit P$, because F is smooth in a neighborhood of $\clubsuit I \cap \clubsuit P$.

The strands of $\heartsuit A^* \cap \clubsuit J_3$ and $\heartsuit A^* \cap \clubsuit J_4$ are C^1 because F is a local diffeomorphism on these pieces, and the strands in the range $\heartsuit A^* \cap (\clubsuit P \cup \clubsuit I)$ are C^1.

For the remaining pieces, those on the top left, $\clubsuit J_i$ and $\clubsuit K_i$ and $\clubsuit L_i$ for $i = 1, 2$, we need a different argument. For each remaining piece we have shown in the preceding lemmas that $dF(0, 1)$ is timelike. We have also shown that the the strands of $\heartsuit A$ are spacelike. In other words, the smooth map F is transverse to each strand of $\heartsuit A$ in the sense of [**GP**, §1, Sect. 5]. The theorem about smooth transversality at the end of this section has the same proof when the target manifolds – in our case the strands of $\heartsuit A$ – are C^1, and in this case the conclusion is that the inverse images of these strands are C^1 arcs.

The only thing left to examine is that happens at the vertical interfaces between the two pieces. In each case, we are using the same map on each side of the interface, and the map extends to be smooth in a neighborhood of the interface. So, the strands of $\heartsuit A^*$ are C^1 even where they cross from one piece to the next. $\square$

16.3. Using Symmetry

Taking a look at Figure 16.1 and comparing the things we proved about $\heartsuit A^*$ we see that all the pieces of $\heartsuit A^*$ fit together to make a union of 2 Cantor cones and a Cantor band, as shown in Figure 16.5. In the figures below we depict the Cantor cones and Cantor products as solid objects. The Cantor product connects the right side of $\clubsuit I_4$ to the left side of $\clubsuit J_4$. The left side of $\clubsuit J_4$ is the same as the right side of the piece $\clubsuit O$ from §13.4.

LEMMA 16.7. *Every point of $\clubsuit P^* - \heartsuit A^*$ except $\boldsymbol{q}$ and $\boldsymbol{r}$ is preconvex.*

Proof: From the work in the previous chapter, all points in $\clubsuit P$ have well defined orbits, and all points in $\clubsuit P - \heartsuit A$ are preconvex. Every point in $p \in \clubsuit P^* - \boldsymbol{q} - \boldsymbol{r}$

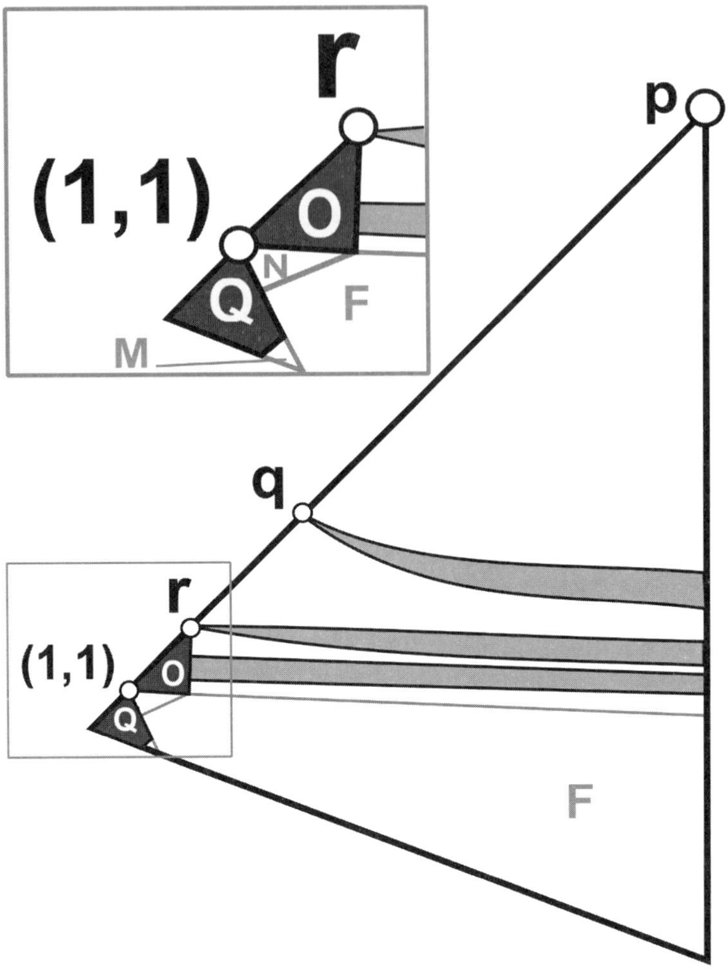

FIGURE 16.5. The set $\heartsuit A^*$

has the property that $F^2(p)$ lies in the union of $\clubsuit P$ and pieces consisting entirely of preconvex points. So, by definition, p is preconvex unless $p \in \heartsuit A^*$. This establishes our claim. □

LEMMA 16.8. *$\heartsuit J - \boldsymbol{D}$ is the $\boldsymbol{\Gamma}$ orbit of $\heartsuit A^* \cup (\heartsuit J \cap \clubsuit O) - \{(1,1)\}$.*

Proof: Our analysis in §13 shows that every point of $\boldsymbol{T} - \clubsuit O - \clubsuit Q - \clubsuit P^*$ is preconvex, except for the points on the diagonal lying in the pre-image of $\mathbf{p}$ and $\boldsymbol{q}$. All these exceptional points lie in $\clubsuit O$. Combining this information with Lemma 16.7, we see that

$$\heartsuit J \cap (\boldsymbol{T} - \clubsuit Q) \subset \heartsuit A^* \cup (\heartsuit J \cap \clubsuit O).$$

The set $\boldsymbol{T} - \clubsuit Q$ is not closed. It omits the two edges of $\clubsuit Q$ lying in the interior of $\boldsymbol{T}$. But all points on these edges lie in $\clubsuit F, \clubsuit M, \clubsuit N$ from §13. From the analysis there, all these points are preconvex except $(1,1)$. So, $\heartsuit J \cap (\boldsymbol{T} - \clubsuit Q)$ also lies in

$\heartsuit A^* \cup (\heartsuit J \cap \clubsuit O) - \{(1,1)\}$. This lemma now follows from the fact that $\boldsymbol{T}$ is a fundamental domain for the action of $\boldsymbol{\Gamma}$. $\square$

Now we will describe the set $\heartsuit A^* \cup (\heartsuit J \cap \clubsuit O) - \{(1,1)\}$. Let h be the restriction of the primary element $\boldsymbol{H}_{101}$ to $\clubsuit O$. In Lemma 13.6, we showed that h is an expanding map on $\clubsuit O$ which fixes the vertex $(1,1)$. Moreover,

$$h(\clubsuit O) \subset \boldsymbol{T} - \clubsuit Q \cup \{(1,1)\}. \tag{16.5}$$

Therefore

$$\text{closure}(\heartsuit J \cap (\boldsymbol{T} - \clubsuit Q)) = \heartsuit A^* \cup (\heartsuit J \cap \clubsuit O) = \{(1,1)\} \cup \bigcup_{k=0}^{\infty} h^{-k}(\heartsuit A^*). \tag{16.6}$$

We set

$$\heartsuit A_k^* = h^{-k}(\heartsuit A^*). \tag{16.7}$$

Now we see how these pieces fit together.

Let $\clubsuit \partial_1 \boldsymbol{O}$ and $\partial_1 \boldsymbol{T}$ respectively denote the vertical edges of $\clubsuit O$ and $\boldsymbol{T}$.

LEMMA 16.9. *$h(\clubsuit \partial_1 \boldsymbol{O}) \subset \partial_1 \boldsymbol{T}$ and $\heartsuit A^* \subset h(\clubsuit O)$.*

Proof: We check by direct calculation that h maps the vertical line $x = \phi^2$ into the vertical line $x = \phi^6$. Since $h(\clubsuit O - \{(1,1)\}) \subset \boldsymbol{T}$, we have $h(\clubsuit \partial_1 \boldsymbol{O}) \subset \partial_1 \boldsymbol{T})$. This establishes the first claim.

For the second claim, we also note that $h(\phi^2, \phi^2) = (\phi^6, \phi^6)$, and that h maps the diagonal edge of $\clubsuit O$ into the diagonal edge of $\boldsymbol{T}$. Finally, we use the confinement test to show that h maps the third edge of $\clubsuit O$ into the union $\clubsuit B \cup \clubsuit F$. These are pieces which lie beneath $\heartsuit A^*$. See Figure 13.5. Hence h maps $\clubsuit O - \boldsymbol{r}$ completely over a region in $\boldsymbol{T}$ which contains $\heartsuit A^*$. $\square$

Note that $\heartsuit A^* \cap \clubsuit \partial_1 \boldsymbol{O}$ and $\heartsuit A^* \cap \partial_1 \boldsymbol{T}$ are both Cantor sets. Since h is a homeomorphism from $\clubsuit \partial_1 \boldsymbol{O}$ into a segment $\sigma \subset \partial_1 \boldsymbol{T}$ which contains $\heartsuit A^* \cap \partial_1 \boldsymbol{T}$ in its interior, and since all other points on $\sigma - \heartsuit A^*$ are preconvex, we must have

$$h(\heartsuit A^* \cap \clubsuit \partial_1 \boldsymbol{O}) = \heartsuit A^* \cap \partial_1 \boldsymbol{T}. \tag{16.8}$$

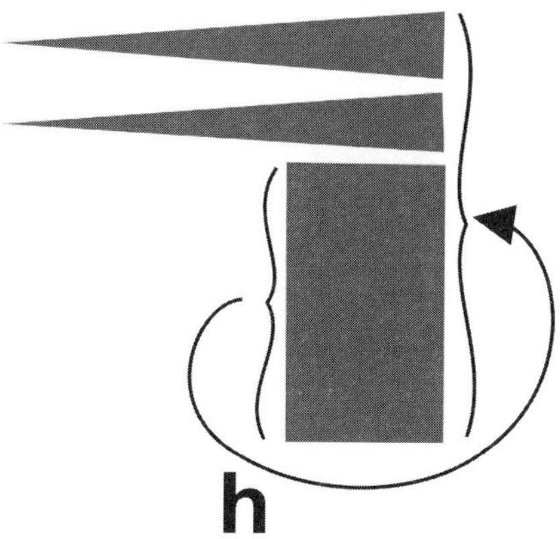

FIGURE 16.6. How h glues the sides of $\heartsuit A^*$

Equation (16.8) tells us that h glues the left side of the Cantor product in $\heartsuit A^*$ to the right side of $\heartsuit A^*$. Figure 16.6 shows what this looks like schematically.

Pulling back by h, and using equation (16.8), we see that the right side of $\heartsuit A_1^*$ coincides with the left side of the Cantor product in $\heartsuit A^*$. Moreover, the strands in the union are C^1 even at the points where they are glued together. The reason is that $\heartsuit J$ contains a C^1 Cantor band $\heartsuit A^{**}$ which contains a neighborhood of the right ends of $\heartsuit A_0^*$, and $h^{-1}(\heartsuit A^{**})$ contains a neighborhood of the points where $\heartsuit A_0^*$ and $\heartsuit A_1^*$ are glued together.

Pulling back by h repeatedly, we get Figure 15.2. We repeat this picture here for convenience.

From this description, we see that the closure of $\heartsuit J \cap (\boldsymbol{T} - \clubsuit Q)$, namely the set in equation (16.6), is a countable union of Cantor cones together with one arc. The limit arc is the union of the point $(1,1)$ and all the bottom arcs of the sets $\heartsuit A_k^*$. Figure 16.7 indicates this bottom arc. Every strand in sight is C^1.

16.4. The Limiting Arc

We have persistently excluded the point $(1,1)$ from consideration. Here we explain one of the difficulties with this point. Even though the strands of $\heartsuit A_k^*$ converge to the limiting arc which runs along the bottom of Figure 16.7, the tangent lines to these arcs along the diagonal do not converge. We will see in the next chapter that the tangent lines to strands of $\heartsuit A_k^*$ converge to the line of slope -1 whereas the tangent line to the limiting arc at $(1,1)$ has slope $+1$. This fact derives

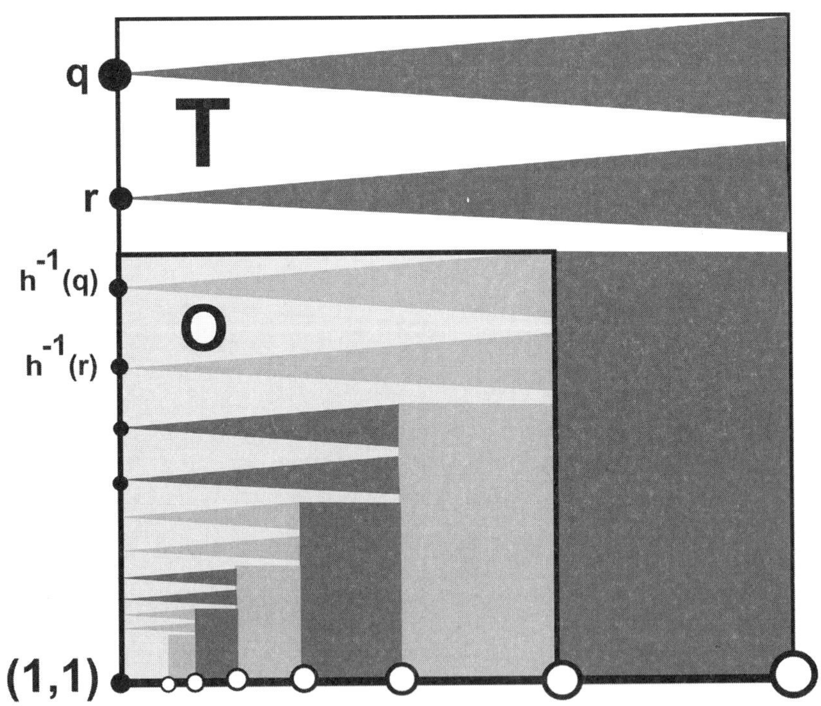

FIGURE 16.7. $\heartsuit A_0^* \cup \cdots \cup \heartsuit A_6^*$ shown schematically

FIGURE 16.8. Near the limiting arc

from the fact that $(1, 1)$ is a fixed point of $\boldsymbol{H}$ and the differential $d\boldsymbol{H}$ the line of slope 1 as its dominant eigen-direction. See §7.4.

Figure 16.8 shows a schematic picture of the strands of $\heartsuit J$ near $(1, 1)$, together with their images under the reflection $\boldsymbol{R}$. The limiting arc and its image under the reflection $\boldsymbol{R}$ meet at a cusp. This is what Figure 16.8 shows. The reader can see a much better picture using the program and zooming into the point $(1, 1)$.

CHAPTER 17

The Solenoid

17.1. Recognizing the BJK Continuum

In this chapter we finish the proof of Theorem 1.10. We continue following the outline given in §15.2. In the last chapter we showed that $\heartsuit J \cap \boldsymbol{T} - \clubsuit Q$ is a union of countably many Cantor cones $\heartsuit S_0, \heartsuit S_1, \heartsuit S_2, \dots$ and a single half-open limiting arc, $\heartsuit S_\infty$. The point $(1,1)$ is the endpoint of $\heartsuit S_\infty$. We note that $\boldsymbol{T} - \boldsymbol{D} = \boldsymbol{T} - \clubsuit Q$. We use these two sets indistinguishably.

Let $\boldsymbol{R}^2_p$ denote the blowup of $\boldsymbol{R}^2$ at a point p. Let $\pi : \boldsymbol{R}^2_p \to \boldsymbol{R}^2$ be the blow-down map. We call $p \in \heartsuit J$ a *basic cone point* if the strands of $\heartsuit J$ have well-defined tangent vectors at p and if the lifted set $\heartsuit J_p$ in $\boldsymbol{R}^2_p$ is a C^1 Cantor band in a neighborhood of the exceptional fiber $\pi^1(p)$.

We say that a line L through p *skewers* the Cantor cone if L is not tangent to any of the strands of $\heartsuit J$ at p. In other words, the lift $\heartsuit J_p$ is disjoint from the point in the exceptional fiber representing L. Below we will prove the following result.

LEMMA 17.1. *for each $j = 0, 1, 2, \dots$ the cone point of $\heartsuit S_j$ is a basic cone point, and $\heartsuit S_j$ is skewered by the line of slope -1.*

Let $\heartsuit \widehat{S}_j$ denote the Cantor product obtained by blowing up the cone point of $\heartsuit S_j$. Let $\heartsuit S$ be the union of all these Cantor products together with the limiting arc. We include the point $(1,1)$ in $\heartsuit S$, and we topologize $\heartsuit S$ so that the limiting arc $\heartsuit \widehat{S}_\infty$ is the limit of the sets $\heartsuit \widehat{S}_j$ as $j \to \infty$.

We identify $\heartsuit S$ with

$$[0,1] \times \boldsymbol{Z}_2. \tag{17.1}$$

So that the point in $\heartsuit \widehat{S}_k$ correspond to those elements of $\boldsymbol{Z}_2$ with sequences of the form $0 \dots 01$, with k initial 0s. The arc $\heartsuit \widehat{S}_\infty$ is $[0,1] \times \{0\}$.

Now we explain how $\boldsymbol{\Gamma}$ acts on our space $\heartsuit S$. Note first that $\boldsymbol{\Gamma}$ acts as a group of diffeomorphisms in a neighborhood of our union $\bigcup \heartsuit S_j$. If we blow up finitely many of the cone points, then the action of $\boldsymbol{\Gamma}$ extends to a group of diffeomorphisms on the resulting space. If we blow up at all the cone points and eliminate the 5-points in the orbit $\boldsymbol{\Gamma}(1,1)$ – i.e. the accumulation set of the blow-up points – then again $\boldsymbol{\Gamma}$ acts as a group of diffeomorphisms. By restriction, $\boldsymbol{\Gamma}$ acts as a group of homeomorphisms on the space $\heartsuit S - \boldsymbol{\Gamma}(1,1)$, the space obtained by deleting the 5 special points from $\heartsuit S$. But the space $\heartsuit S$ is just the 5-point compactification of this "punctured space" and the action of $\boldsymbol{\Gamma}$ naturally extends to a group of homeomorphisms on $\heartsuit S$.

Now we analyze the quotient. $\heartsuit S/\boldsymbol{\Gamma}$ and then we take the 10-fold cover. As in Figure 10.3, let $\partial_0 \boldsymbol{T}$ and $\partial_1 \boldsymbol{T}$ respectively denote the diagonal and vertical sides of $\boldsymbol{T}$. See Figure 10.3.

Recall the two involutions on $\boldsymbol{Z}_2$ defined in §6.7. The involution I_0 reverses all the digits after the first 1 is encountered. For instance,

$$I_0(\mathbf{001}0101110\ldots) = \mathbf{001}1010001\ldots$$

The involution I_1 reverses all the digits.

LEMMA 17.2. *The action of $\boldsymbol{R}$ on $\partial_0 \boldsymbol{T}$ induces the action of the involution I_0 from §6.7 on $\{0\} \times \boldsymbol{Z}_2$.*

Proof: The element $\boldsymbol{R}$ fixes $\partial_0 \boldsymbol{T}$ pointwise and acts as a reflection. Hence $\boldsymbol{R}$ fixes the cone point of $\heartsuit S_k$ and reverses the ordering on the strands of $\heartsuit S_k$. That is, $\boldsymbol{R}$ maps the tangent line to the top strand to the tangent line of the bottom strand, and so on. Crucially, what makes this work is that, by Lemma 17.1, the line of slope -1 through the cone point, which is preserved by $\boldsymbol{R}$, skewers $\heartsuit S_k$. So, the involution does not fix any of the tangent lines to the strands. This shows that the action of on the left endpoints of strands of $\heartsuit \widehat{S}_k$ induced by $\boldsymbol{R}$ is the same as the action of I_0 on the corresponding subset of $\{0\} \times \boldsymbol{Z}_2$. Hence the action of $\boldsymbol{R}$ on $\partial_0 \boldsymbol{T}$ induces the action on $\heartsuit S \cap \partial_0 \boldsymbol{T}$ corresponding to the action of I_0 on $\{0\} \times \boldsymbol{Z}_2$. □

LEMMA 17.3. *The action of $\boldsymbol{H}_{130}$ on $\partial_1 \boldsymbol{T}$ induces the action of the involution I_1 from §6.7 on $\{1\} \times \boldsymbol{Z}_2$.*

Proof: A direct calculation, which we omit, shows that $\boldsymbol{H}_{130}$ stabilizes the vertical edge $\partial_1 \boldsymbol{T}$ and reverses this edge, fixing the point (ϕ^6, ϕ^2). This is the white dot shown on the right boundary component of $\clubsuit G$ in Figure 13.2. (Use the pointer function of the computer program for a graphic demonstration of this.) Let $\heartsuit \Sigma_0$ denote the union of the right endpoints of $\heartsuit S_0$. Let $\heartsuit \Sigma_1$ denote the union of the right endpoints of the remaining Cantor bands, and the right endpoint of $\heartsuit S_\infty$. The set $\heartsuit \Sigma_0$ lies above the piece $\clubsuit G$ and $\heartsuit \Sigma_0$ lies below $\clubsuit G$. Hence $\heartsuit \Sigma_0$ lies above (ϕ^6, ϕ^2) and $\heartsuit \Sigma_1$ lies below (ϕ^6, ϕ^2). Again, see Figure 13.2.

Since the only points in $\partial_1 \boldsymbol{T}$ in $\heartsuit J$ are those in the Cantor cone A^*, we see that $\boldsymbol{H}_{130}$ an involution of a Cantor band neighborhood of $\heartsuit \Sigma_0 \cup \heartsuit \Sigma_1$ and this involution swaps $\heartsuit \Sigma_0$ with $\heartsuit \Sigma_1$, reversing the top to bottom order. Hence, the involution induced by $\boldsymbol{H}_{130}$ on $\heartsuit S \cap \partial_1 \boldsymbol{T}$ corresponds to the action of I_1 on $\{1\} \times \boldsymbol{Z}_2$. □

From these two results, we recognize $\heartsuit S/\boldsymbol{\Gamma}$ as the BJK continuum described in §6.7.

17.2. Taking Covers

We continue with the notation from the previous section. We have just shown that $\heartsuit S/\boldsymbol{\Gamma}$ is the BJK continuum. Hence $\heartsuit S$ is a 10-fold cover of the BJK continuum. Looking at Figure 17.1, which shows $\heartsuit S$ schematically, we see that $\heartsuit S$ is actually the 5-fold cover of the 2-adic solenoid (as opposed to some other 10-fold cover of the BJK continuum.) This completes the proof of Theorem 1.10, modulo the analysis of the cone points.

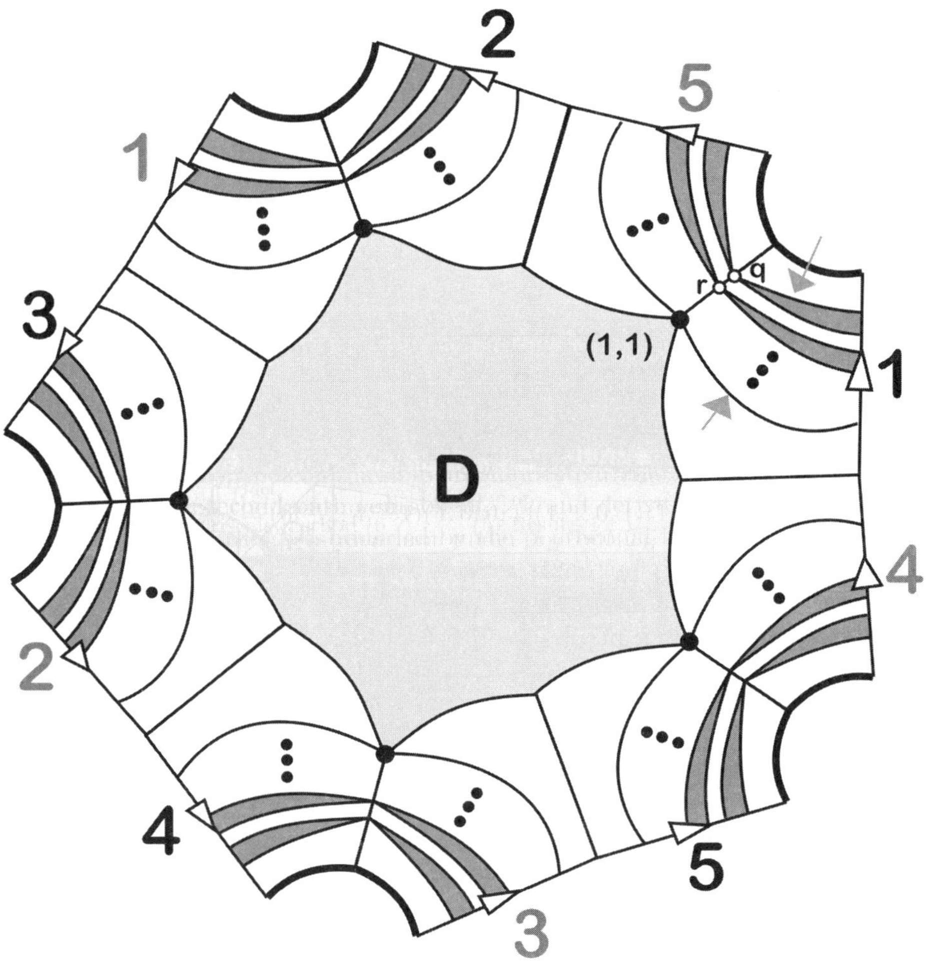

FIGURE 17.1. The 10-fold cover

17.3. Connectivity and Unboundedness

Before we move on to the analysis of the cone points there is more we want to say more about the structure of $\heartsuit J - \boldsymbol{D}$. These properties will be important in our proof that $\heartsuit J$ is path connected. The following result is a precursor to our theorem that $\heartsuit J$ is path connected.

LEMMA 17.4. *$\heartsuit J - \boldsymbol{D}$ is path connected.*

Proof: Let S denote the 5-fold cover of the 2-adic solenoid. Each maximal curve in S is dense in S. Moreover, at most one direction of such a maximal curve can encounter one of the special points corresponding to the vertices of the disk $\boldsymbol{D}$. For this reason, if we start with a point of $\heartsuit J - D$ we can follow it around in a C^1 way, through the cone points, until it lands in the Cantor cone $\heartsuit S_0$. But then we can continue until we hit the cone point of $\heartsuit S_0$. Hence, every point of $\heartsuit J - \boldsymbol{D}$ can be connected to the point $\boldsymbol{q}$, which is the cone point of $\heartsuit S_0$. $\square$

LEMMA 17.5. *Every infinite C^1 path in $\heartsuit J - \boldsymbol{D}$ exits $\boldsymbol{R}^2$. More strongly, a C^1 path in the closure of $\heartsuit J - \boldsymbol{D}$ cannot go through* 11 *consecutive tiles in the set $\Gamma(\boldsymbol{T})$ without exiting $\boldsymbol{R}^2$.*

Proof: The element $\boldsymbol{H}_{020}$ maps $\boldsymbol{R} \times \{\infty\}$ to a connected curve which contains the top vertex (ϕ^6, ϕ^6) of $\boldsymbol{T}$ and also a point on the bottom edge of $\boldsymbol{T}$. Call this arc α. Note that $\boldsymbol{H}_{030} = \boldsymbol{H}_{020}^{-1}$. Every C^1 arc of $\heartsuit J - \boldsymbol{C}$ which cuts across the fundamental domain $\boldsymbol{T}$ (i.e. enters on one side and leaves on another) must cross α. Hence, any C^1 arc of $\heartsuit J - \boldsymbol{C}$ which cuts across $\boldsymbol{H}_{030}(\boldsymbol{T})$ must exit $\boldsymbol{R}^2$. Any infinite C^1 path of $\heartsuit J - \boldsymbol{D}$ must cut across $\boldsymbol{H}_{030}(\boldsymbol{T})$, and indeed any C^1 path of $\heartsuit J - \boldsymbol{D}$ which enters 11 consecutive tiles in $\boldsymbol{\Gamma}(\boldsymbol{T})$ must cut across every tile in $\boldsymbol{\Gamma}(\boldsymbol{T})$, including $\boldsymbol{H}_{030}(\boldsymbol{T})$. □

17.4. The Canonical Loop

In this section we will isolate a particular closed loop in $\heartsuit J - \boldsymbol{D}$. This loop will be important in the proof of Theorem 1.9. Even though the cover of the solenoid has no closed loop, when we blow it down to $\heartsuit J - \boldsymbol{D}$ we get a closed loop. Our closed loop is not C^1, however. There are no C^1 loops in $\heartsuit J - \boldsymbol{D}$.

The loop we have in mind is the $\boldsymbol{\Gamma}$ orbit of two arcs in $\boldsymbol{T}$. One of the arcs connects $(1, 1)$ to the vertical side of $\boldsymbol{T}$. This is the arc we have been calling *the limiting arc*. The other arc connects the point $\boldsymbol{q}$ to the vertical side of $\boldsymbol{T}$. These two arcs have little arrows pointing to them in Figure 17.1.

The $\boldsymbol{\Gamma}$ orbit of these two arcs is a topological circle which is C^1 except at 10 points, namely the the $\boldsymbol{\Gamma}$ orbit of $\boldsymbol{q}$ and the vertices of $\boldsymbol{D}$. The 5 cone points and the 5 vertices alternate. The path has a "kink" at each cone point, and a "cusp" at each vertex point. We call this loop *the canonical loop*.

17.5. Using Symmetry for the Cone Points

To finish the proof of Theorem 1.10 we prove Lemma 17.1. In this section we reduce Lemma 17.1 to a simpler result. The cone points of our Cantor cones cone in two infinite families.

$$h^{-k}(\boldsymbol{p}), \qquad k = 1, 2, 3, \ldots \qquad\qquad h^{-k}(\boldsymbol{q}), \qquad k = 0, 1, 2, \ldots \tag{17.2}$$

Here

- $\boldsymbol{p} = (\phi^6, \phi^6)$ is a vertex of the triangle $\boldsymbol{T}$. We call $\boldsymbol{p}$ the *apex* of $\boldsymbol{T}$. This is one of the 3 points we blow up to get the space $\boldsymbol{M}$.
- $\boldsymbol{q}$ has a complicated formula, but $\boldsymbol{q} = \boldsymbol{B}(\psi, \psi)$, where $\psi = (1 + \sqrt{13})/2$.
- $\mathbf{r} = (\phi^2, \phi^2)$.
- The map h is the restriction of the primary element $\boldsymbol{H}_{101} = \boldsymbol{RH}$ to the piece P in the partition.

Lemma 17.1 is a consequence of the following result.

LEMMA 17.6. *The point $\mathbf{p}$ is a basic cone point for $\pi(\heartsuit J)$ and the point $\boldsymbol{q}$ is a basic cone point for $\heartsuit J$. The lines of slope -1 through $\mathbf{p}$ and $\boldsymbol{q}$ are not tangent to any of the strands of $\pi(\heartsuit J)$ or $\heartsuit J$ respectively.*

Here $\pi : \boldsymbol{M} \to (\boldsymbol{R} \cup \infty)^2$ is the blow-down map. The map h is an expanding diffeomorphism in a neighborhood of the piece P, which contains the remaining cone points. Moreover $h(\boldsymbol{r}) = \boldsymbol{p}$, and the remaining cone points are the pre-images of $\boldsymbol{q}$ and $\boldsymbol{r}$ under j. Hence, all these points are basic cone points.

Since h commutes with the reflection $\boldsymbol{R}$ in the diagonal, the differential dh maps lines of slope -1 to lines of slope -1. So, the line of slope -1 is never tangent to one of the strands of our Cantor cones at a cone point. This proves Lemma 17.1 The rest of the chapter is devoted to proving Lemma 17.6.

17.6. The First Cone Point

In this section we analyze $\pi(\heartsuit J)$ in a neighborhood of $\boldsymbol{p}$. We will first study how $\heartsuit J$ looks in a neighborhood of the exceptional fiber $\pi^{-1}(\boldsymbol{p})$ and then we will blow down and see how $\pi(\heartsuit J)$ looks in a neighborhood of $\boldsymbol{p}$. We will flip back and forth between the original coordinates and the $\boldsymbol{B}$-coordinates.

We start with the original coordinates. Recall that M is the manifold on which the Gauss group Γ acts. M is obtained from $(\boldsymbol{R} \cup \infty)$ by blowing up at $(1,1)$, $(\infty, 0)$ and $(0, \infty)$. From the formula

$$H_{140}(x,y) = \left(x, \frac{y-1}{xy-1}\right), \tag{17.3}$$

we see that H_{140} maps the exceptional fiber $\pi^{-1}(1,1)$ to the vertical line $x = 1$. Applying $\boldsymbol{B}$ we see that $\boldsymbol{H}_{140}$ maps the exceptional fiber $\pi^{-1}(\boldsymbol{p})$ to the vertical line $\boldsymbol{L}$ which is the lift of $x = \phi^6$. Note that this vertical line contains an exceptional point on $\pi^{-1}(\boldsymbol{p})$. In short, there is a diffeomorphism of $\boldsymbol{M}$ which carries the intersection of $\heartsuit J$ with a neighborhood of $\pi^{-1}(\boldsymbol{p})$ to the intersection of $\heartsuit J$ with a neighborhood of $\boldsymbol{L}$. Figure 17.2 shows how (the finite points of) a neighborhood of L projects into $\boldsymbol{R}^2$. Figure 17.2 also shows a line X whose lift to M misses this neighborhood entirely.

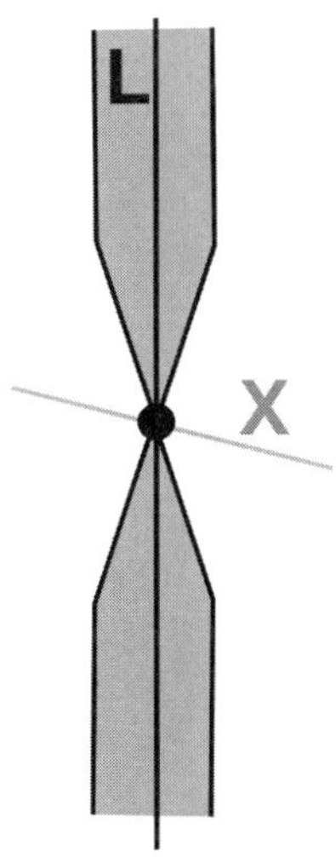

FIGURE 17.2. A neighborhood of the line L and a line X whose lift misses it

LEMMA 17.7. *$\boldsymbol{L}$ intersects $\heartsuit J$ transversely in a Cantor set consisting entirely of finite points.*

Proof: $\boldsymbol{L}$ is the lift of the line extending the vertical edge $\partial_1 \boldsymbol{T}$ of $\boldsymbol{T}$. In the previous chapter we showed that $\heartsuit J \cap \partial_1 \boldsymbol{T} - \{\boldsymbol{p}\}$ is precisely the Cantor set of right endpoints of A^*. We also know that this intersection is transverse, by Lemma 16.1.

The set $\boldsymbol{L} - \boldsymbol{T}$ is simply a boundary of the space $\boldsymbol{C}$ of convex classes and hence does not intersect $\heartsuit J$. See Lemma 9.8.

It remains to see how $\boldsymbol{L} \cap \pi^{-1}(\boldsymbol{p})$ intersects $\heartsuit J$. As we observed in the remark following the proof of Lemma 9.8, the map H is entirely defined on the line $x = 1$ and maps the point $(1,1)$ into $(0,1)^2$, the set of points representing the convex classes. Hence $\boldsymbol{H}$ maps the exceptional point on $\boldsymbol{L}$ into the space $\boldsymbol{C} = \boldsymbol{B}((0,1)^2)$. In other words, the exceptional point of $\boldsymbol{L}$ is not in $\heartsuit J$. $\square$

It follows from transversality that a neighborhood of $\boldsymbol{L}$ in $\boldsymbol{M}$ intersects $\heartsuit J$ transversely in a Cantor band. Hence a neighborhood of $\pi^{-1}(\boldsymbol{p})$ intersects $\heartsuit J$ in a Cantor band which is transverse to $\pi^{-1}(\boldsymbol{p})$. Applying the blow-down map, we see that $\boldsymbol{p}$ is a basic cone point for the set $\pi^{-1}(\heartsuit J)$.

LEMMA 17.8. *The point of $\pi^{-1}(\boldsymbol{p})$ corresponding to the line of slope -1 is not a point in $\heartsuit J$. In other words, the line of slope -1 through $\mathbf{p}$ is not tangent to any arc of $\pi(\heartsuit J)$.*

Proof: Let $\boldsymbol{p}_{-1}$ denote the point in $\pi^{-1}(p)$ corresponding to the line of slope -1. We have $\boldsymbol{H}_{140}(\boldsymbol{p}_{-1}) = (\phi^6, \phi^2)$. This point lies outside the region $\clubsuit P^*$ from the previous chapter. In particular, it does not lie in A^*. Hence the line of slope -1 through $\boldsymbol{p}$ is not tangent to $\pi(\heartsuit J)$. $\square$

This proves the first half of Lemma 17.6.

17.7. The Second Cone Point

Now we study the structure of $\heartsuit J$ in a neighborhood of $\boldsymbol{q}$. Of the pieces of our partition having $\boldsymbol{q}$ as a vertex, the only ones containing points which are not preconvex are $\clubsuit L_1$ and $\clubsuit L_2$. Therefore, in a neighborhood of $\boldsymbol{q}$ the only points of $\heartsuit J$ are contained in $\clubsuit L \cup \boldsymbol{R}(\clubsuit L)$.

$$h_1 = \boldsymbol{H}_{111}, \qquad h_2 = \boldsymbol{H}_{041}. \tag{17.4}$$

These are the primary maps we associated to the "strips" $\clubsuit P_1^*$ and $\clubsuit P_2^*$ in the previous chapter. Note that $\clubsuit L_1 \subset \clubsuit P_1^*$ and $\clubsuit L_2 \subset \clubsuit P_2^*$.

In Lemma 16.5 we analyzed the behavior of h_j on $\clubsuit L_j - \boldsymbol{q}$, or rather we explained that the situation was the same as in Lemma 16.4. That is, h_j maps each vertical segment of $\clubsuit L_j$ into a timelike curve which runs transverse to the Cantor band $\heartsuit A$ and remains inside $\boldsymbol{T} \cup \spadesuit C$. Recall that $\spadesuit C$ does not intersect $\heartsuit J$. Hence $\heartsuit J = h_j^{-1}(\heartsuit A)$ inside $\clubsuit L_j$.

LEMMA 17.9. *Each coordinate function of h_j has a simple blowup at $\boldsymbol{q}$.*

Proof: Consider $h_1 = \boldsymbol{H}_{111}$ first. Recall that $\psi = (1 + \sqrt{13})/2$ and $\boldsymbol{q} = \boldsymbol{B}(\psi, \psi)$. Since $\boldsymbol{B}$ is a diffeomorphism and (ψ, ψ) is not a blow-up point for the Gauss group, the statement that the coordinate functions of $\boldsymbol{H}_{111}$ have simple blowups at $\boldsymbol{q}$ is the same as the statement that the coordinate functions of H_{111} have simple blowups at (ψ, ψ). This is a direct calculation, which we omit. The proof for h_2 is similar. $\square$

Let $\boldsymbol{R}^2_{\boldsymbol{q}}$ be the blowup of $\boldsymbol{R}^2$ at $\boldsymbol{q}$ and let π be the blow-down map. h_j lifts to a smooth map $\widehat{h}_j : \widehat{S} \to \boldsymbol{R}^2$. Here $\widehat{S} = \pi^{-1}(S)$, where S is the intersection of a sufficiently small disk with $\clubsuit L \cup \boldsymbol{R}(\clubsuit L)$. The reason why the range is $\boldsymbol{R}^2$ is that the image $h_j(\clubsuit L_j - \boldsymbol{q})$ is uniformly bounded.

The vertical line foliation of S lifts to a smooth foliation of $\widehat{S}$ in which one of the leaves λ is a subset of the exceptional fiber. $\widehat{h}_j : \heartsuit \widehat{S}_j \to \boldsymbol{R}^2$ maps each curve, except perhaps λ, to a timelike curve. Moreover, the restriction of $\widehat{h}_j$ to λ is a regular map because h_j has a simple blowup at $\boldsymbol{q}$. So, by continuity, $\widehat{h}_j(\lambda)$ is not spacelike. Hence $\widehat{h}_j$ maps every leaf in the foliation to a smooth regular curve that is transverse to $\heartsuit A$.

The same argument as in Lemmas 16.4 and 16.6 shows that $\widehat{h}_j^{-1}(\heartsuit A)$ is a C^1 Cantor band transverse to λ. Since this holds for each j, there is a C^1 Cantor band in $\widehat{S}$ whose blowdown is $\heartsuit J \cap S$. This proves that $\boldsymbol{q}$ is a basic cone point.

LEMMA 17.10. *The line of slope -1 through $\boldsymbol{q}$ is not tangent to any of the strands of $\heartsuit J$.*

Proof: Let $\boldsymbol{q}_{-1}$ denote the point in the exceptional fiber of $\boldsymbol{R}^2_{\boldsymbol{q}}$ corresponding to the line of slope -1. We compute that $\widehat{h}_1(\boldsymbol{q}_{-1})$ and $\widehat{h}_2(\boldsymbol{q}_{-1})$ both lie in the bottom edge of $\boldsymbol{T}$. They are the same point, namely

$$(\phi^2 t, -t), \qquad t = \frac{6 + 4\sqrt{5} - 4\sqrt{13} - 2\sqrt{65}}{-9 - 5\sqrt{5} + \sqrt{13} + \sqrt{65}} \tag{17.5}$$

No point on the bottom edge of $\boldsymbol{T}$ outside of $\clubsuit Q$ belongs to $\heartsuit J$. This completes the proof. $\square$

CHAPTER 18

Local Structure of the Julia Set

18.1. Blowing Down the Exceptional Fibers

In this chapter we prove Theorem 1.8. We follow the sketch given in §15.3. Again $\heartsuit J$ is the Julia set and $\boldsymbol{D}$ is the big disk used in the proof of Theorem 1.5. Let $\pi : \boldsymbol{M} \to (\boldsymbol{R} \cup \infty)^2$ be the blow-down map. We persistently confuse the disks $\boldsymbol{D}$ and $\pi^{-1}(\boldsymbol{D})$ because $\boldsymbol{D}$ is disjoint from the blow-up points. $\heartsuit J$ as a subset of $\boldsymbol{M}$.

We already know that every point of $\heartsuit J - \boldsymbol{D}$ is either a band point or a basic cone point. All the cone points here are $\boldsymbol{\Gamma}$ images of the ones covered by Lemma 17.1. In particular, every point of $\heartsuit J \cap \pi^{-1}(\boldsymbol{R}^2) - \boldsymbol{D}$ is either a band point or a basic cone point. We would like to see that every point of $\heartsuit \pi(J) \cap \boldsymbol{R}^2 - \boldsymbol{D}$ is either a cone point or a band point. Note that $\heartsuit \pi(J) \cap \boldsymbol{R}^2$ and $\heartsuit J \cap \boldsymbol{R}^2$ coincide except at the blow-up points. We have already seen that the blow-up point (ϕ^6, ϕ^6) is a cone point. We just have to check that the two other blow-up points $(-\phi^4, \phi^2)$ and $(\phi^2, -\phi^4)$. are either cone points or band points. By symmetry it suffices to check the first of these two points. We will show that $a = (-\phi^4, \phi^2)$ is a cone point.

The element $\boldsymbol{H}_{030}$ blows a up to the line λ_1 given by the equation $y = \phi^2$. Let $\lambda_2 \subset \boldsymbol{T}$ denote the set of points which are $\boldsymbol{\Gamma}$ equivalent to λ_1. It turns out that

$$\lambda_2 = (\lambda_1 \cup \boldsymbol{R}(\lambda_1)) \cap \boldsymbol{T}. \tag{18.1}$$

The reason for this is as follows:

(1) The element $\boldsymbol{H}_{100} = \boldsymbol{R}$ preserves $\lambda_1 \cup \boldsymbol{R}(\lambda_1)$ and, around their point of intersection, locally swaps the vertical and horizontal segments of λ_2.

(2) The element $\boldsymbol{H}_{120}$ preserves $\boldsymbol{R}(\lambda_1)$, fixes the bottom endpoint of the vertical segment of λ_2, and reverses the direction.

(3) The element $\boldsymbol{H}_{130}$ preserves λ_1, fixes the right endpoint of the horizontal segment of λ_2, and reverses the direction.

One might say that the action of $\boldsymbol{\Gamma}$ folds $\lambda_1 \cup \boldsymbol{R}(\lambda_1)$ onto λ_2. This fact depends on the special nature of λ_1; it does not work this way for other horizontal lines. Figure 18.1 below shows a picture of λ_2. The bottom half of $\boldsymbol{T}$ is also shown, as are approximate drawings of the partition pieces $\clubsuit F, \clubsuit G, \clubsuit J_3, \clubsuit J_4, \clubsuit O$.

The set λ_2 is exactly the set of interfaces between the shaded regions in Figure 10.3. Let $C\lambda_2$ denote the corner of λ_2. This is the point (ϕ^2, ϕ^2). Let $H\lambda_2$ denote the horizontal segment of λ_2 minus $C\lambda_2$. Let $V\lambda_2$ denote the vertical segment of λ_2 minus $C\lambda_2$.

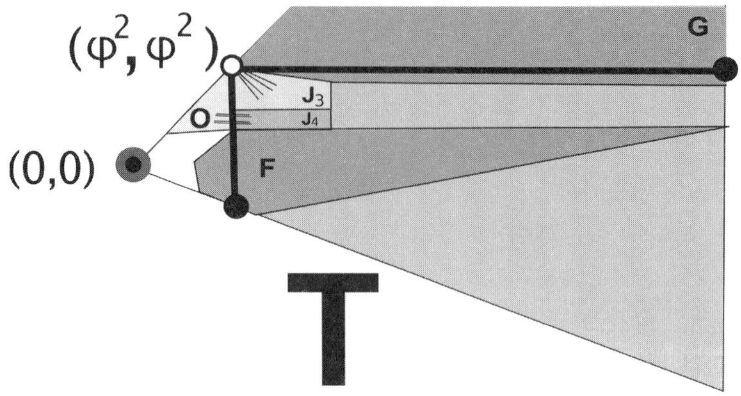

FIGURE 18.1. The set λ_2

We consider these three subsets of λ_2 in turn.

(1) The segment $H\lambda_2$ lies in $\clubsuit G$, a piece which does not intersect $\heartsuit J$.

(2) The segment $V\lambda_2$ lies in the same line as the vertical edge $\clubsuit O \cap \clubsuit J_4$, and the segment of $V\lambda_2$ below $\clubsuit O \cap \clubsuit J_4$ lies in $\clubsuit F$, a piece disjoint from $\heartsuit J$. Hence $V\lambda_2$ intersects $\heartsuit J$ transversely in the Cantor set which is the set of right endpoints of the Cantor band $\boldsymbol{A}_1^*$ from §16.1.

(3) $C\lambda_2$ is a cone point, as we have seen in the last chapter. The strands of $\heartsuit J$ emanating from this cone point all lie in $\clubsuit J_3$ and hence have negative slope. This is to say that neither segment of λ_2 is tangent to the strands of $\heartsuit J$ at this point.

The projection from λ_1 to λ_2 is a 5-fold orbifold cover, folding at the two far endpoints of $H\lambda_2$ and $V\lambda_2$. Hence, from the analysis above, we conclude that $\pi^{-1}(-\phi^4, \phi^2)$ is transverse to 5 Cantor bands $b_1, \ldots, b_5$ and 5 cone points $c_1, \ldots, c_5$. The sets $\pi(b_i)$ and $\pi(c_j)$ are each Cantor cones individually. Moreover, since all these sets are disjoint along the exceptional fiber, the set of tangent lines to one of the projections is completely disjoint from the set of tangent lines to any of the others. Hence, the union $\bigcup \pi(b_i) \cup \bigcup \pi(c_i)$ is still a Cantor cone.

Hence $(-\phi^4, \phi^2)$ is a cone point for $\pi(\heartsuit J) \cap \boldsymbol{R}^2$. Now we know that every point of $\pi(\heartsuit J) \cap \boldsymbol{R}^2 - \boldsymbol{D}$ is either a band point or a cone point.

Before we move on to the next section we observe that the primary map $\boldsymbol{H}_{011}$ blows up the point

$$\mathbf{t} = (\phi^{-2}, \phi^{-2}) \tag{18.2}$$

to the same line $y = \phi^2$ that we have just considered. Hence $\mathbf{t}$ is also a cone point. On the other hand, not all the primary maps blow up at $\mathbf{t}$. For instance, the map $\boldsymbol{H}_{031}$ does not. The point $\mathbf{t}$ is shown in Figure 12.4.

18.2. Everything but One Piece

Now we will consider what happens in $\heartsuit J \cap \boldsymbol{D} - \heartsuit JC$. Since all the blow-up points lie outside $\boldsymbol{D}$ we will frequently confuse the two sets $\heartsuit J \cap \boldsymbol{D}$ and $\pi(\heartsuit J) \cap \boldsymbol{D}$.

In §12 we produced a union $\boldsymbol{D}_1 \cup \cdots \cup \boldsymbol{D}_6$ of 6 topological disks together with primary maps $h_j : \boldsymbol{D}_j \to \boldsymbol{R}^2$ which map $\boldsymbol{D}_j$ over $\boldsymbol{D}$ and which act as a diffeomorphism on $\boldsymbol{D}_j$.

LEMMA 18.1. *Theorem 1.8 is true provided that every point of*

$$\heartsuit J \cap (\boldsymbol{D} - \boldsymbol{D}_1 - \cdots - \boldsymbol{D}_6)$$

is a cone point or a band point.

Proof: To establish Theorem 1.8 we have to prove every point of $\heartsuit J \cap \boldsymbol{D} - \heartsuit JC$ is a cone point or a band point. Let $\boldsymbol{D}^* = \boldsymbol{D}_1 \cup \cdots \cup \boldsymbol{D}_6$. Let $F : \boldsymbol{D}^* \to \boldsymbol{R}^2$ be the map which agrees with h_j on $\boldsymbol{D}_j$. Every point $p \in \boldsymbol{D}^* - \heartsuit JC$ has the property that there is some m such that $F^m(a) \in \boldsymbol{R}^2 - \boldsymbol{D}^*$. But F^m is a local diffeomorphism at a. If we knew that every point of $\pi(\heartsuit J) \cap (\boldsymbol{R}^2 - \boldsymbol{D}^*)$ is either a band point or a cone point, then we know that a is a band point or a cone point. We already know that $\pi(\heartsuit J) \cap (\boldsymbol{R}^2 - \boldsymbol{D})$ is a cone point or a band point. This just leaves points in $\pi(\heartsuit J) \cap (\boldsymbol{D} - \boldsymbol{D}^*)$, and $\heartsuit J = \pi(\heartsuit J)$ inside $\boldsymbol{D}$. □

Figure 18.2 repeats the picture of the partition used in the proof of Theorem 1.5.

We have $\boldsymbol{D} - (\boldsymbol{D}_1 \cup \cdots \cup \boldsymbol{D}_6) \subset \boldsymbol{\Gamma}(\clubsuit S)$. By symmetry, we just have to show that each point of $\heartsuit J \cap \clubsuit S$ Is a cone point or a band point. Recall that $\clubsuit S = \clubsuit S_1 \cup \clubsuit S_2 \cup \clubsuit S_3$.

LEMMA 18.2. *Each point of $\heartsuit J \cap \clubsuit S_2$ is either a cone point or a band point.*

Proof: Let $h = \boldsymbol{H}_{141}$. Let $\clubsuit S_2' = \clubsuit S_2 - \mathbf{s}$. We have already seen in our proof of Theorem 1.5 that h is defined and finite on $\clubsuit S_2'$. We use the area test to check that $\det(dh) - \phi^{-2} \geq 0$ on $\clubsuit S_2'$. Hence h is a local diffeomorphism on $\clubsuit S_2'$. We also saw that $h(\clubsuit S_2') \subset \boldsymbol{R}^2 - \boldsymbol{D}_0$. Hence every point of $\heartsuit J \cap \clubsuit S_2$ is either a cone point or a band point. □

LEMMA 18.3. *Each point of $\heartsuit J \cap \clubsuit S_3$ is either cone point or a band point.*

Proof: Let $h = \boldsymbol{H}_{031}$. We use the denominator test to check that h is defined and finite on $\clubsuit S_3$. We already know that some primary element maps $\clubsuit S_3$ outside $\boldsymbol{D}_0$ and by symmetry so does h. In short, $h(\clubsuit S_3) \subset \boldsymbol{R}^2 - \boldsymbol{D}_0$. We use the area test to show that $\det(dh) > 0$ on $\clubsuit S_3$. Hence h is a local diffeomorphism on $\clubsuit S_3$. □

18.3. The Last Piece

It remains to show that each point in $\heartsuit J \cap \clubsuit S_1$ is either a cone point or a band point. None of the primary maps is a local diffeomorphism on all of $\clubsuit S_1$. We partition $\clubsuit S_1$ into 7 smaller pieces $\clubsuit S_{11}, \ldots, \clubsuit S_{17}$ as shown in Figure 18.3. The label k denotes $\clubsuit S_{1k}$. The polygons are shaded according to the role they play

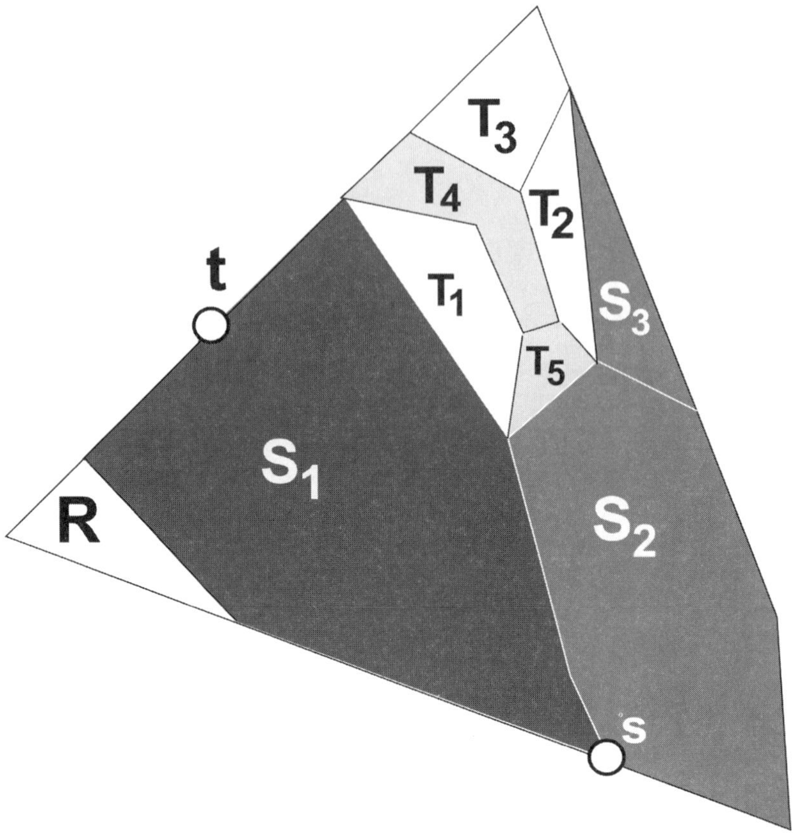

FIGURE 18.2. The partition of $\clubsuit Q$

in the proof. We list the vertices in §21.7. We have already seen that the point $\mathbf{t} = (\phi^{-2}, \phi^{-2})$ is a basic cone point.

We assign primary elements to these pieces as follows:

(1) $h_1 = \boldsymbol{H}_{011}$.
(2) $h_2 = \boldsymbol{H}_{111}$.
(3) $h_3 = \boldsymbol{H}_{131}$.
(4) $h_4 = \boldsymbol{H}_{111}$.
(5) $h_5 = \boldsymbol{H}_{011}$.
(6) $h_6 = \boldsymbol{H}_{011}$.
(7) $h_7 = \boldsymbol{H}_{111}$.

Let $\clubsuit S'_{1j} = \clubsuit S_{1j} - \mathbf{s} - \mathbf{t}$. We check for each j that h_j is defined and finite on $\clubsuit S'_j$. In all cases (whether we need to or not) we use the weak denominator test coupled with the subtraction trick.

18.3.1. The Local Diffeomorphisms.

LEMMA 18.4. *Each point of* $\clubsuit S_{11} \cup \clubsuit S_{12} \cup \clubsuit S_{13}$ *except possibly for* $\mathbf{s}$ *is a cone point or a band point.*

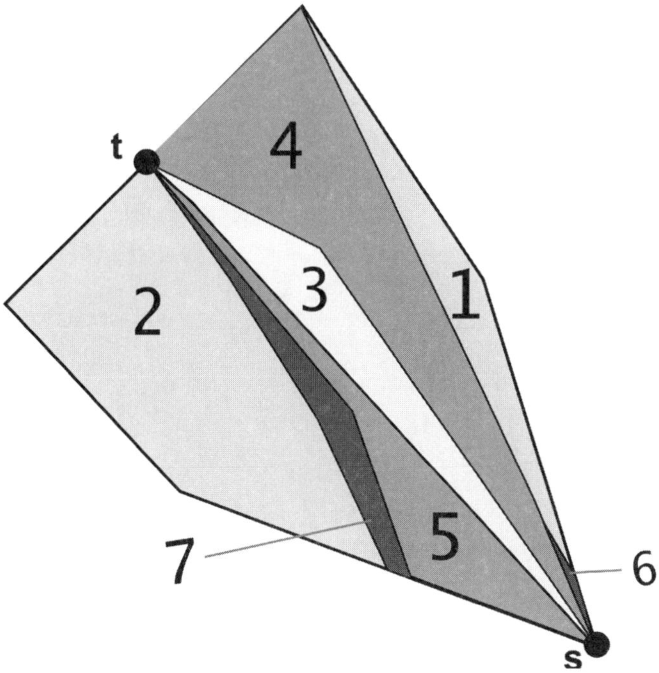

FIGURE 18.3. The pieces $\clubsuit S_{11}, \ldots, \clubsuit S_{17}$

Proof: We use the strong area test to show that $\det(dh_1) > 0$ on $\clubsuit S_{11}$. We use the weak area test to show that $\det(dh_2) \geq \phi^{-2}$ on $\clubsuit S_2'$. For $\clubsuit S_3'$ be use the weak area test combined with the variation trick to show that $\det(dh_1) > 0$ on $\clubsuit S_3'$. Hence h_j is a local diffeomorphism on $\clubsuit S_j'$ for $j = 1, 2, 3$. The rest of the proof is as above. $\square$

Remark: For later purposes we also check that $h_3(\clubsuit S_{13} - \mathbf{s})$ does not contain one of the 3 points in $\boldsymbol{R}^2$ which we blew up to create $\boldsymbol{M}$. We use the confinement test to show that $h_3(\clubsuit S_{13} - \mathbf{s})$ is contained in the interior of the rectangle $[0, \phi^2] \times [-10, 0]$. This rectangle is disjoint from two of the blow-up points, and has the blow-up point $(\phi^2, -\phi^4)$ on its right edge.

18.3.2. Avoided.

LEMMA 18.5. *For $j = 4, 5$, the set $\clubsuit S_{1j} - \mathbf{s} - \mathbf{t}$ is disjoint from $\heartsuit J$.*

Proof: We use the confinement test to check that

$$h_4(\clubsuit S_4') \subset \clubsuit B \cup \clubsuit F \cup \clubsuit N \cup \spadesuit C \cup \spadesuit E \tag{18.3}$$

The union on the right is convex. Figure 13.3 shows these pieces. We have shown in §13 that these partition pieces are disjoint from $\heartsuit J$.

Let $\spadesuit Y_1$ denote the rectangle of width 10 whose sides are parallel to the coordinate axis and whose left side coincides with the right side of $\clubsuit G$. Let $\spadesuit Y_2 = \clubsuit G \cup \spadesuit Y_1$. We are just prolonging $\clubsuit G$ somewhat. We use the confinement test to check that $h_5(\clubsuit S_{15}) \subset \spadesuit Y_2$. So, it suffices to prove that $\spadesuit Y_2 \cap \heartsuit J = \emptyset$. We use the confinement test to show that

$$\boldsymbol{H}_{021}(\spadesuit Y_1) \subset \spadesuit D.$$

We already know that $\clubsuit G \cap \heartsuit J = \emptyset$, and the last calculation tells us $\spadesuit Y_1 \cap \heartsuit J = \emptyset$. Hence $\spadesuit Y_2 \cap \heartsuit J = \emptyset$. □

18.3.3. First Painful Piece. Now we come to the first of two painful pieces. To save words in our proofs, we remark here that we use the denominator test to check that every map in sight is defined on the relevant domain.

LEMMA 18.6. *Every point of $\heartsuit J \cap \clubsuit S_{16} - \mathbf{s}$ is a band point.*

Proof: We introduce auxiliary quadrilaterals $\spadesuit Y$ and $\spadesuit Z$ whose vertices we list in §21.8. Figure 18.4 shows $\spadesuit Y$ and how $\heartsuit J$ sits inside of it. We get $\spadesuit Z$ by slightly enlarging $\boldsymbol{H}_{101}(\spadesuit Y)$. We check that

$$\boldsymbol{H}_{101}(\spadesuit Y) \subset \spadesuit Z, \tag{18.4}$$

and

$$\spadesuit Z \subset \clubsuit P_3 \cup \clubsuit P_4 \cup \clubsuit B \cup \clubsuit F \cup \clubsuit G \cup \boldsymbol{R}(\clubsuit G) - \boldsymbol{q}. \tag{18.5}$$

The pieces $\clubsuit P_3$ and $\clubsuit P_4$ are the bottom two rows of the quasi-horseshoe $\clubsuit P$. The pieces $\clubsuit F$ and $\clubsuit G - \boldsymbol{q}$ are disjoint from $\heartsuit J$. By symmetry $\boldsymbol{R}(\clubsuit G) - \boldsymbol{q}$ is also disjoint from $\heartsuit J$. From all this we conclude that $\boldsymbol{H}_{101}(Y) \cap \heartsuit J \subset \boldsymbol{A}$, where $\boldsymbol{A}$ is the Cantor band from the proof of Theorem 1.6. Hence $h_6(\clubsuit S_{16} - \mathbf{s} \cap \heartsuit J)$ consists entirely of band points.

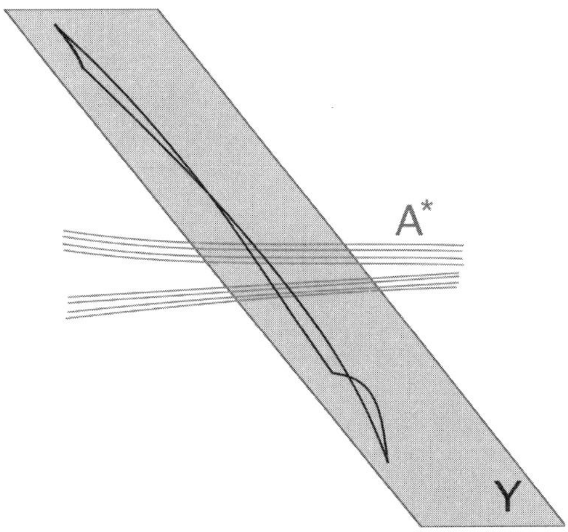

FIGURE 18.4. The image $h_6(\partial \clubsuit S_{16})$

We foliate A_{16} (minus its top and bottom vertex) by horizontal line segments. We know that the strands of $\boldsymbol{A}$ are spacelike. That is, their tangent lines never lie in the standard lightcone $\vee = \vee(1, -1)$.

We check the following facts.

- $dh_6(1, 0) \subset \vee(-\infty, -1/3)$ at all points in $\clubsuit S_{16} - \mathbf{s}$.
- $d\boldsymbol{H}_{101}(\vee(-\infty, -1/3)) \subset \vee(-1/4, 4)$ at all points of $\spadesuit Y$.

- Let h_3 and h_4 be the primary elements associated to $\clubsuit P_3$ and $\clubsuit P_4$. Then $dh_j(\vee(-1/4,4)) \subset \vee$ throughout $\clubsuit P_j$.

We use the cone test in all cases.

These results imply that there is a semigroup element h so that $h(\clubsuit S_{16} - \mathbf{s})$ intersects $\heartsuit J$ only in the Cantor band $\boldsymbol{A}$, and h maps the horizontal foliation to timelike curves. Since the strands of $\boldsymbol{A}$ are spacelike, we see that h maps the strands of the horizontal foliation transverse to $\boldsymbol{A}$. This proves what we want. □

18.3.4. Second Painful Piece. Before we deal with the last piece, we prove a preliminary result. Referring to the partition piece $\clubsuit J_3$ used in the proof of Theorem 1.6, we write

$$\clubsuit J_3 = \clubsuit J_{31} \cup \clubsuit J_{32}, \tag{18.6}$$

We introduce 2 small polygons $\spadesuit Y_1$ and $\spadesuit Y_2$. Both these polygons are symmetric with respect to reflection in the diagonal. Two of the vertices of $\spadesuit Y_1$ are the bottom left and top right corners of $\clubsuit J_3$. This determines $\spadesuit Y_1$. One edge of $\spadesuit Y_2$ is the left edge of $\clubsuit J_3$. This determines $\spadesuit Y_2$.

LEMMA 18.7.

$$h_7(\heartsuit J \cap \clubsuit S_{17} - \mathbf{t}) \subset \bigcup_{i=0}^{1} \boldsymbol{R}^i(\heartsuit J \cap \clubsuit J_{31}). \tag{18.7}$$

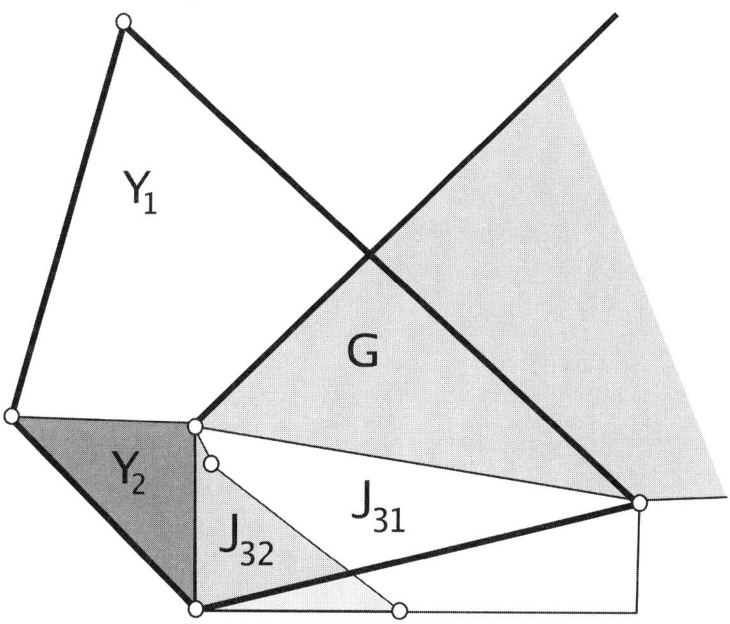

FIGURE 18.5. The pieces $\clubsuit Y_1, \clubsuit Y_2, \clubsuit J_{31}, \clubsuit J_{32}, \clubsuit G$

Proof: We check that

$$\spadesuit Y_1 \subset \bigcup_{i=0}^{1} \boldsymbol{R}^i(\spadesuit Y_2 \cup \clubsuit J_{31} \cup \clubsuit J_{32} \cup \clubsuit G). \tag{18.8}$$

We use the confinement test to show that

$$h_7(\clubsuit S'_{17}) \subset \spadesuit Y_1. \tag{18.9}$$

Next we check that $\boldsymbol{H}_{101}(\spadesuit Y_2) \subset \clubsuit A$. Since $\clubsuit A \cap \heartsuit J = \emptyset$, we have $\heartsuit J \cap \spadesuit Y_2 = \emptyset$. By symmetry $\heartsuit J \cap \boldsymbol{R}(\spadesuit Y_2) = \emptyset$. We check that

$$\boldsymbol{H}_{141}(\clubsuit J_{32}) \subset \clubsuit B \cup \clubsuit F \cup \spadesuit C \cup \spadesuit E. \tag{18.10}$$

None of these partition pieces intersects $\heartsuit J$, so we conclude that $\clubsuit J_{32} \cap \heartsuit J = \emptyset$. By symmetry $\boldsymbol{R}(\clubsuit J_{32}) \cap \heartsuit J = \emptyset$. Putting all this information together gives us equation (18.7). □

Now for the last piece.

LEMMA 18.8. *Each point of $\heartsuit J \cap \clubsuit S_{17}$ is either a cone point or a band point.*

Proof: We already know that $\heartsuit J \cap \clubsuit J_3$ is a Cantor cone. Hence the right hand side of equation (18.7) is the union of two Cantor cones meeting at the cone point $\boldsymbol{q}$. We foliate A_{17} by horizontal line segments above, and as in the previous result we show that h_7 maps these segments transverse to $\heartsuit J$.

Here are the details of the transversality argument. We use the cone test to show that h_7 maps these horizontal segments to curves of positive slope.

On the other hand, we use the cone test to check that $d\boldsymbol{H}_{141}(\vee(0,\infty)) \subset \vee$ throughout $\clubsuit J_{31}$. Next, we check

$$\boldsymbol{H}_{141}(\clubsuit J_{31} \cap \heartsuit J) \subset \clubsuit I \cup \clubsuit P. \tag{18.11}$$

In view of the fact that all the strands of $\heartsuit J \cap (\clubsuit I \cup \clubsuit P)$ are spacelike, we conclude that the strands of $\heartsuit J$ have negative slope in $\clubsuit J_{31}$. By symmetry, the same holds for the strands of $\heartsuit J$ in $\boldsymbol{R}(\clubsuit J_{31})$.

In conclusion, h_7 maps the horizontal segments to curves of positive slope whereas all the strands of $\heartsuit J$ they encounter have negative slope. This gives us the transversality we need. □

Before we leave this section, we prove one more result about $\clubsuit S_{17}$ which will be useful when we prove that $\heartsuit J$ is connected.

LEMMA 18.9. *$\heartsuit J$ does not intersect the boundary of $\clubsuit S_{17}$ except at the point* $\mathbf{t}$ *or in the edge of $\clubsuit S_{17}$ contained in the bottom edge of $\boldsymbol{T}$.*

Proof: Some of the edges of $\clubsuit S_{17}$ are also edges of the adjacent tile $\clubsuit S_{15}$, and we know that $\clubsuit S_{15} \cap \heartsuit J = \mathbf{t}$. So, Ψ cannot intersect any of these edges. For each remaining edge e of $\clubsuit S_{17}$, other than the bottom edge, we use the confinement test to check that $\boldsymbol{H}_{111}(e) \subset \spadesuit Y_2$, a piece we have already shown to be disjoint from $\heartsuit J$. □

18.4. The Last Point

The only points of $\clubsuit S \cap \heartsuit J$ we have not dealt with is the point $\mathbf{s}$. In this section we show that $\mathbf{s}$ is a basic cone point. Our proof, unfortunately, is also rather painful. We follow the same argument we used in §18.1, but here we have a more complicated image to worry about.

A calculation shows that the two elements $\boldsymbol{H}_{031}$ and $\boldsymbol{H}_{141}$ both blow up $\mathbf{s}$ to the horizontal line λ_3, given by the equation $y = y_0$, where

$$y_0 = \frac{(-73 - 120\phi) + (35 + 55\phi)\psi)}{8} \approx 2.29518. \tag{18.12}$$

Figure 18.6 shows λ_3. To finish the proof we have to show that λ_3 intersects $\heartsuit J$ transversely in a Cantor set. Our strategy is the same as what we did for the blow-up point $(-\phi^4, \phi^2)$ above. We will take the $\boldsymbol{\Gamma}$ orbit of γ_3 and see how this set intersects $\heartsuit J \cap \boldsymbol{T}$.

Without giving many details of the calculation, which we do using Mathematica, we explain how it is done. We compute that the image $\lambda_3' = \boldsymbol{B}^{-1}(\lambda_3)$ is the horizontal line $y = y_0'$, where

$$y_0' = -6 - 5\psi, \qquad \psi = \frac{1 + \sqrt{13}}{2}. \tag{18.13}$$

We compute the Γ orbit of λ' and then apply $\boldsymbol{B}$.

We find that the $\boldsymbol{\Gamma}$ orbit of λ_3 consists of the horizontal line λ_3, the vertical line $\lambda_4 = \boldsymbol{R}(\lambda_3)$ and two hyperbolas, λ_5 and λ_6. Both hyperbolas have horizontal and vertical asymptotes, and these asymptotes intersect on the main diagonal. We just have to understand how these 4 curves intersect $\mathcal{J}$ inside the fundamental domain $\boldsymbol{T}$. We will consider the 4 curves in turn.

18.4.1. The Horizontal and Vertical Lines. Figure 18.6 shows λ_3 and λ_4 as well as the surrounding partition pieces. (The pieces $\clubsuit F$, $\clubsuit G$, and $\clubsuit P$ are not fully shown.)

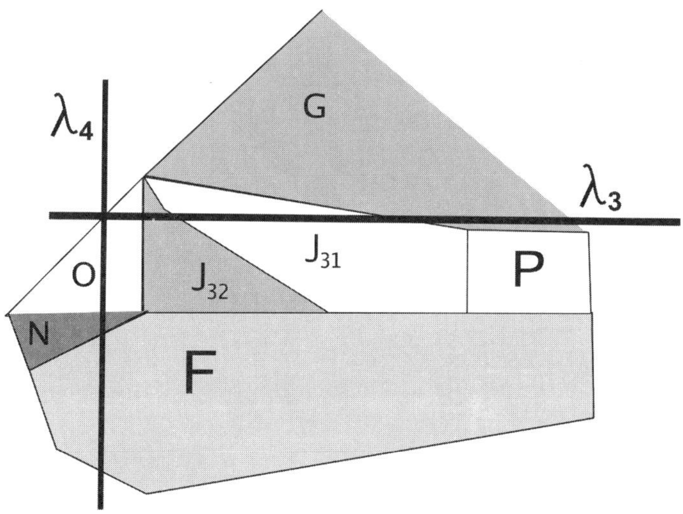

FIGURE 18.6. λ_3 and λ_4 and the relevant pieces

For λ_3, the only relevant pieces in the partition which intersect $\heartsuit J$ non-trivially are $\clubsuit O$ and J_{31}. We use the confinement test to check that

$$\boldsymbol{H}_{101}(\lambda_3 \cap \clubsuit O) \subset \clubsuit A.$$

Since $\clubsuit A \cap \heartsuit J = \emptyset$, we see that $\lambda_3 \cap \clubsuit O \cap \heartsuit J = \emptyset$.

We have already seen in the previous section that the strands of $\heartsuit J$ in $\clubsuit J_{31}$ have negative slope. Hence λ_3 is transverse to $\heartsuit J$ in $\clubsuit J_{31}$. Hence $\lambda_3 \cap \boldsymbol{T}$ intersects $\heartsuit J$ transversely in a Cantor set.

Now we consider λ_4. Let $h = \boldsymbol{H}_{101}$. We already know that $h(\clubsuit O) \subset \boldsymbol{T}$. We use the confinement test to check that $h(\lambda_4 \cap \clubsuit O)$ lies to the right of the vertical line through the rightmost point of $\clubsuit P$. The only pieces in the partition to the right of this $\clubsuit P$ containing $\heartsuit J$ are those of $\clubsuit I$. Hence

$$h(\lambda_4 \cap \heartsuit J) \subset \clubsuit I. \tag{18.14}$$

We use the cone test to show that $dh(0,1) \subset \vee$ on $\lambda_4 \cap \clubsuit O$. Since all the strands of $\heartsuit J \cap (I \cup \clubsuit P)$. are spacelike, we conclude that λ_4 is transverse to $\heartsuit J$.

18.4.2. The First Hyperbola. Figure 18.7 shows an accurate picture of part of λ_5. This hyperbola intersects $\boldsymbol{T}$ in the vertex (ϕ^6, ϕ^6) and a point (t, t) where $t \approx -.87$. In particular, λ_5 only intersects $\boldsymbol{T}$ at the vertex (ϕ^6, ϕ^6).

By symmetry, the tangent line to λ_5 at (ϕ^6, ϕ^6) has slope -1. Since the group $\boldsymbol{\Gamma}$ acts on the space $\boldsymbol{M}$ obtained by blowing up 3 points, including (ϕ^6, ϕ^6), we need to consider the image of λ_5 in $\boldsymbol{M}$ to figure out how the corresponding point of λ_5 intersects $\heartsuit J$. We have already observed that the point in $\boldsymbol{M}$ corresponding to the line of slope -1 through (ϕ^6, ϕ^6) does not intersect $\heartsuit J$. Hence, the corresponding point of λ_5 does not intersect $\heartsuit J$. So, the corresponding point in the fiber above $\mathbf{s}$ does not intersect $\heartsuit J$. In short, λ_5 contributes nothing to the structure of $\heartsuit J$ near $\mathbf{s}$. (Again, by construction we only need to understand how our curves intersect $\mathcal{J}$ inside the fundamental domain $\boldsymbol{T}$.)

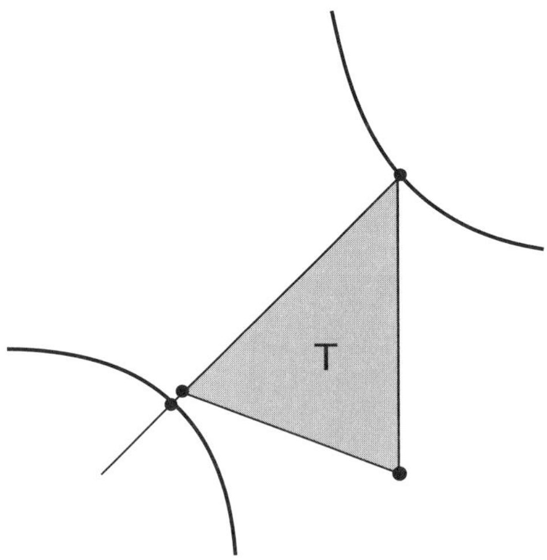

FIGURE 18.7. λ_5 and $\boldsymbol{T}$

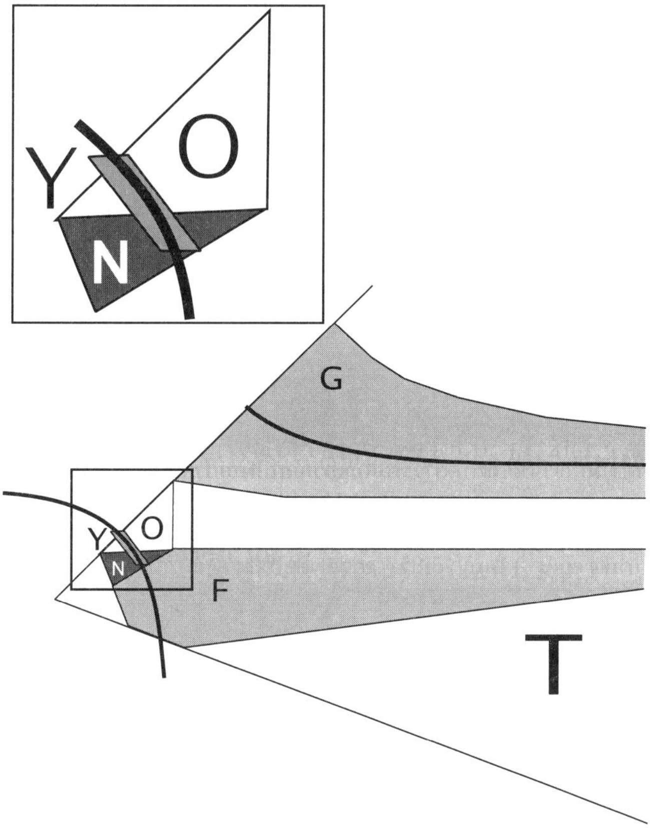

FIGURE 18.8. λ_6 and $\boldsymbol{T}$

18.4.3. The Second Hyperbola. The hyperbola λ_6 has a messy equation, but $\lambda_6 = \boldsymbol{B}(\lambda_6')$ where λ_6' has the equation

$$xy = \frac{19 + 5\sqrt{13}}{2}. \tag{18.15}$$

Figure 18.8 shows a fairly accurate picture. λ_6 intersects $\boldsymbol{T}$ in two connected components, an upper branch and a lower branch. We first show that the upper branch contributes nothing.

LEMMA 18.10. *One component of $\lambda_6 \cap \boldsymbol{T}$ lies entirely in $\clubsuit G$. Hence this component does not intersect $\heartsuit J$.*

Proof: The upper component of $\lambda_6 \cap \boldsymbol{T}$ is the graph of a convex decreasing function. Hence, it is contained in the triangle whose vertices are p_1, p_2, p_3. Here p_1 and p_2 are respectively the left and right endpoints of the branch, and p_3 is the intersection of the vertical line through p_1 and the horizontal line through p_2. We check that $\clubsuit G$ contains this triangle. $\square$

Now we consider the lower branch of $\lambda_6 \cap \boldsymbol{T}$. Let $\spadesuit Y$ and $\spadesuit Z$ be the quadrilaterals from Lemma 18.6. A calculation shows that some segment of λ_6 is contained in $\spadesuit Y$ and connects a point on the top edge of $\spadesuit Y$ to a point on the bottom edge of $\spadesuit Y$. Moreover, λ_6 intersects the top edge of $\spadesuit Y$ in a point that lies outside $\boldsymbol{T}$. Also, the portion of $\lambda_6 \cap \boldsymbol{T}$ below $\spadesuit Y$ lies in the union $\clubsuit N \cup \clubsuit F$. These partition pieces are disjoint from $\heartsuit J$. From all this we conclude that

$$\lambda_6 \cap \heartsuit J \cap \boldsymbol{T} \subset \spadesuit Y \cap \boldsymbol{T}. \tag{18.16}$$

The lines tangent to $\lambda_6 \cap \spadesuit Y \cap \boldsymbol{T}$ lie in $\vee(-\infty, -1)$ because all such points lie below the diagonal. We use the cone test to check

$$d\boldsymbol{H}_{101}(\vee(-\infty, -1)) \subset \vee(1/2, -1/2) \tag{18.17}$$

on the interior of $\clubsuit N \cup \clubsuit O$, a set which contains $\spadesuit Y \cap \cap \boldsymbol{T}$.

Recall from Lemma 18.6 that

$$h(\spadesuit Y) \subset \spadesuit Z, \qquad \spadesuit Z \cap \heartsuit J \subset \clubsuit I \cup \clubsuit P. \tag{18.18}$$

By Lemma 16.1, all strands of $\heartsuit J \cap (\clubsuit I \cap \clubsuit P)$ have slopes in the interval $(-1/2, 1/2)$. So, we see that $\boldsymbol{H}_{101}$ maps $\lambda_6 \cap \spadesuit Y \cap \boldsymbol{T}$ transversely to $\heartsuit J$. Hence $\lambda_6 \cap \spadesuit Y$ is transverse to $\heartsuit J$.

This completes the unpleasant proof that $\mathbf{s}$ is a cone point. Our proof of Theorem 1.8 is done.

18.5. Some Definedness Results

Here we state and prove some lemmas which are essentially just summaries of the analysis done above. These lemmas will be useful in the next chapter.

LEMMA 18.11. *Suppose that $\alpha \in \heartsuit J \cap \boldsymbol{D}$ is some band point. When we consider the range of the primary maps to be the manifold $\boldsymbol{M}$, every primary element is defined at α and maps α to a band point.*

Proof: Our analysis above shows that there is at least one primary element h such that $h(\alpha) \subset \boldsymbol{R}^2$ and $h(\alpha)$ is also a band point of $\heartsuit J$. But then, and other primary element is the composition of h with some diffeomorphism of $\boldsymbol{M}$ which preserves $\heartsuit J$. $\square$

LEMMA 18.12. *Suppose that α is a cone point of $\heartsuit J$ that lies in the interior of $\boldsymbol{D} \cap \boldsymbol{T}$. Then there is some primary element h such that h is a well defined map from a neighborhood of α into $\boldsymbol{R}^2$ and $h(\alpha)$ is a cone point of $\pi(\heartsuit J)$.*

Proof: This is really just a summary of our analysis above. It follows immediately from what we did. $\square$

CHAPTER 19

The Embedded Graph

19.1. Defining the Generator

In this chapter we prove Theorem 1.9. We follow the outline given in §15.4. Our first goal is to find the generator $\heartsuit X$, which is a certain connected arc in $\heartsuit J \cap \boldsymbol{D}$. Our discussion of $\heartsuit X$ refers to Figure 19.1, which is essentially a copy of Figure 17.1.

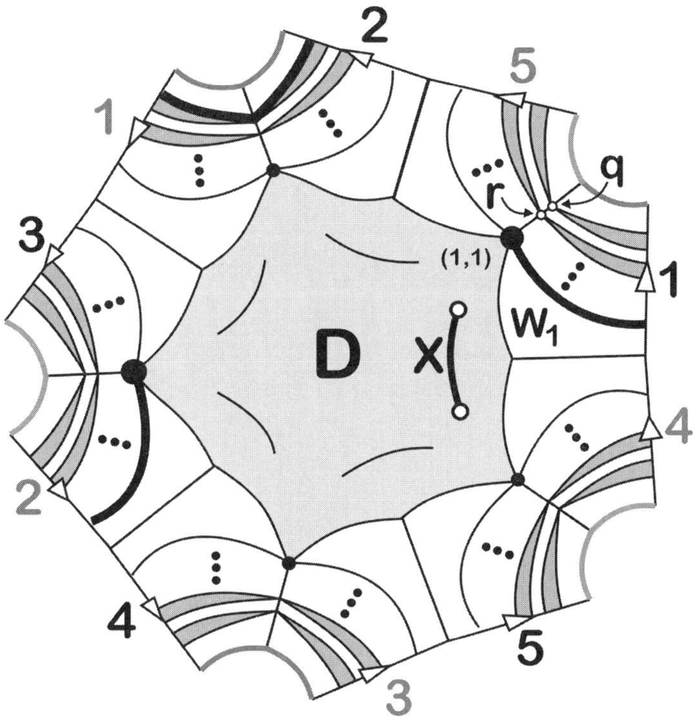

FIGURE 19.1. The arc $\heartsuit W_1$ sitting inside $\heartsuit J - \boldsymbol{D}$

Let $\heartsuit W$ denote the canonical loop described in §17.4. We can view $\heartsuit W$ as an embedded pentagon whose 5 vertices are the 5 irregular fixed points of $\boldsymbol{H}$. These are the vertices of $\boldsymbol{D}$, the $\boldsymbol{\Gamma}$ orbit of $(1,1)$. Let $\heartsuit W_1 \subset \heartsuit W$ be the arc of $\heartsuit W$ which connects two consecutive irregular fixed points, in the cyclic order imposed by $\heartsuit W$. In other words $\heartsuit W_1$ is one fifth of $\heartsuit W$. To pin $\heartsuit W_1$ down exactly, we insist that one of the fixed points is $(1,1)$ and that $\heartsuit W_1$ starts out by running along the bottom arc of $\heartsuit J \cap \boldsymbol{T} - \boldsymbol{D}$. In Figure 19.1, $\heartsuit W_1$ appears to be the union of

4 thick arcs, but the endpoints of these arcs are identified by the gluings to that $\heartsuit W_1$ is a single segment. Figure 19.2 shows $\heartsuit W_1$ schematically.

Recall from §12.5 that we found disks $\boldsymbol{K}_1, \ldots, \boldsymbol{K}_6$ such that each $\boldsymbol{K}_i$ mapped diffeomorphically onto the big disk $\boldsymbol{D}$. We indexed these disks a certain way in §12.5 but there was nothing special about the way we had done it. We specify the indices of these disks now, perhaps in a different way, to suit our present purposes.

Now we discuss Figure 19.2. Let $\boldsymbol{K}_1$ be the disk which has the vertex $(1,1)$. Figure 19.2 shows $\boldsymbol{K}_1$ and also our (new) choices for $\boldsymbol{K}_2$ and $\boldsymbol{K}_3$. The fundamental domain $\boldsymbol{T}$ is shaded. The point a_1 is the vertex of $\boldsymbol{K}_1$ adjacent to $(1,1)$ and lying in $\boldsymbol{T} \cap \boldsymbol{D}$. Recall that $\partial_2 \boldsymbol{T}$ is the non-diagonal non-vertical edge of $\boldsymbol{T}$. (See Figure 10.3.) We draw this edge horizontally in Figure 19.2 and label it with a 2. This edge is fixed pointwise by the element $\gamma = \boldsymbol{H}_{120}$. The point $\gamma(a_1)$ is labeled a_2.

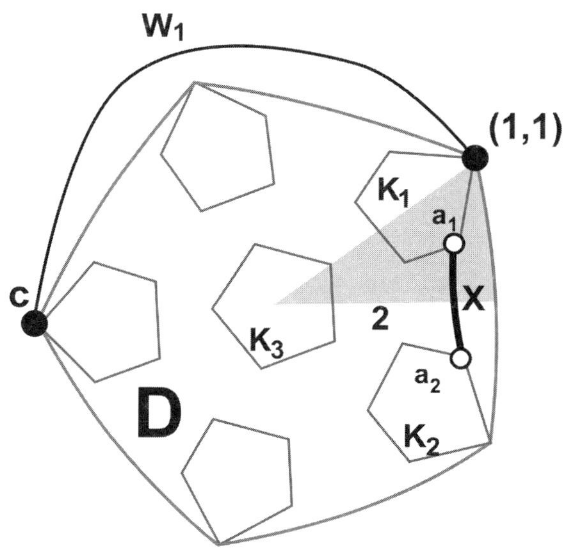

FIGURE 19.2. The sets $\heartsuit W_1$ and $\heartsuit X$ and other relevant points

We want to show that there is an arc $\heartsuit X$ which connects a_1 to $\gamma(a_1)$ and has the following properties.

- $h(\heartsuit X) = \heartsuit W_1$, where $h = \boldsymbol{H}_{031}$.
- $\gamma(\heartsuit X) = \heartsuit X$, where $\gamma = \boldsymbol{H}_{120}$.
- $\heartsuit X$ is disjoint $\boldsymbol{K}_3 \cup \boldsymbol{K}_4 \cup \boldsymbol{K}_5 \cup \boldsymbol{K}_5$.
- $\heartsuit X$ intersects $\boldsymbol{K}_1$ only at a_1.
- $\heartsuit X$ intersects $\boldsymbol{K}_2$ only at a_2.

We will prove this result through a number of small steps.

We have $h(a_1) = (1,1)$ and h is a diffeomorphism in a neighborhood of a_1. Hence, some subset of $h^{-1}(\heartsuit W_1)$ starts at a_1 and moves away from a_1 along a Cantor band. Let $\heartsuit I$ be the closure of the maximal arc like this, whose interior consists entirely of band points. By Lemma 18.11, the map h is defined on the interior of $\heartsuit I$ and h maps the interior of $\heartsuit I$ into $\heartsuit W_1$.

LEMMA 19.1. *$\heartsuit I$ cannot contain a point outside $\boldsymbol{D}$.*

Proof: Every point on $\partial \boldsymbol{D} \cap \boldsymbol{T}$ lies in the $\clubsuit F \cup \clubsuit M \cup \clubsuit N$, and these pieces intersect $\heartsuit J$ only at $(1,1)$. Hence $\heartsuit J$ is disjoint from $\partial \boldsymbol{D}$ except at the 5 vertices. But then $\heartsuit I$ has to exit $\boldsymbol{D}$ at a vertex. But these vertices are not band points. □

LEMMA 19.2. *$\heartsuit I$ cannot stay inside $\boldsymbol{T} \cap \boldsymbol{D}$ and connect two points of $\heartsuit JC$.*

Proof: We will suppose this happens and derive a contradiction. We note first that the only points of $\heartsuit W_1$ in $\heartsuit JC$ are vertices of $\boldsymbol{D}$, and these two vertices are separated by a single cone point on $\heartsuit W_1$. We also recall that $\heartsuit W_1 = \heartsuit h(X)$. Applying Lemma 18.11, we see that $\heartsuit W_1$ would connect 2 vertices of $\boldsymbol{D}$ without intersecting a cone point. Given what we have already noted about $\heartsuit W_1$, this situation is only possible $h(\heartsuit I)$ doubles back on itself before crossing the cone point. That is, h maps both endpoint of $\heartsuit I$ to the same point. We rule this out.

$\heartsuit I$ cannot connect a_1 to itself because then h would not be a local diffeomorphism in a neighborhood of a_1. What other points of $\heartsuit JC$ can $\heartsuit I$ reach? Figure 19.3 shows the various possibilities. (Some of the possibilities shown in Figure 19.3 will be explained in later lemmas in this section.)

Because primary elements map the boundaries of the $\boldsymbol{K}$-disks to the boundary of $\boldsymbol{D}$, we see that $\heartsuit J$ is disjoint from each of the $\boldsymbol{K}$ disk boundaries except at the vertices. So, $\heartsuit I$ cannot enter any of the $\boldsymbol{K}$-disks except at a vertex. There are only 2 vertices of $\boldsymbol{K}$-disks in $\boldsymbol{D} \cap \boldsymbol{T}$. These are the points a_3 and a_5. So, in this situation, $\heartsuit I$ either connects a_1 to a_3 or a_5. We check that $h(a_3) = h(a_5) = d \neq h(a_1)$. This is a contradiction. □

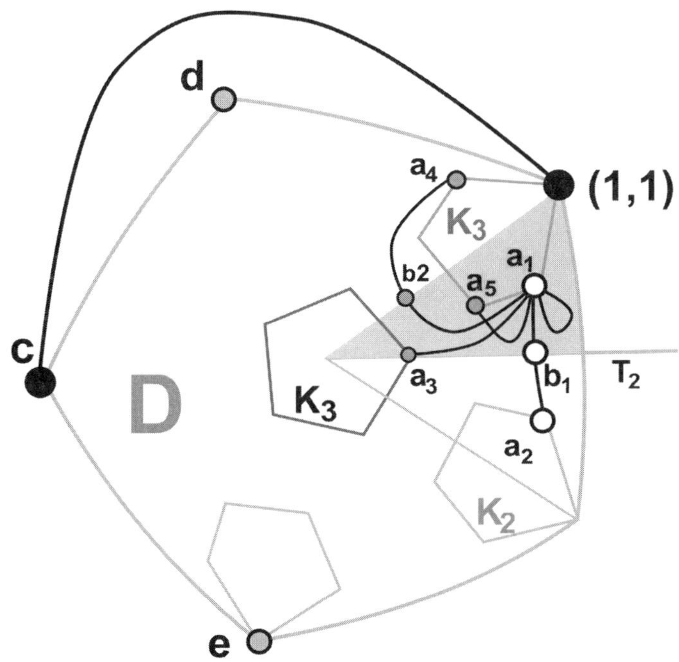

FIGURE 19.3. The possibilities for $\heartsuit X$.

LEMMA 19.3. *An interior point of $\heartsuit I$ cannot be on the boundary of $\boldsymbol{T}$.*

Proof: If this happens, then one of the two edges e of $\boldsymbol{T}$ incident to the origin contains an interior point $\alpha \in \heartsuit I$. There is an involution $\gamma \in \boldsymbol{\Gamma}$ which fixes e pointwise. By Lemma 18.11, we know that $h(\alpha)$ is a band point of $\heartsuit W_1$. On the other hand, $h(\alpha)$ is a point of $\heartsuit W_1$ stabilized by an element of $\boldsymbol{\Gamma}$. The only point on $\heartsuit W_1$ stabilized by an element of $\boldsymbol{\Gamma}$ is the cone point. This is a contradiction. □

LEMMA 19.4. *$\heartsuit I$ cannot lie entirely in the interior of $\boldsymbol{T}$.*

Proof: Suppose that the endpoint α of $\heartsuit I$ is a cone point contained in the interior of $\boldsymbol{T}$. By Lemma 18.12 we know that there is some primary element h' such that $h'(\alpha)$ is a cone point of $\heartsuit\pi(W) \cap \boldsymbol{R}^2$. Here π is the blow-down map. But none of the cone points of $\heartsuit\pi(W)$ is a blow-up point in $\boldsymbol{R}^2$. Indeed, these points are the orbit of the point $\boldsymbol{q}$ in Figure 19.1. The coordinates of points in this orbit do not lie in $\boldsymbol{Z}[\phi]$ and hence are not any of the blow-up points for the manifold $\boldsymbol{M}$. (The 3 blow-up points are listed in §15.1.) So, $h'(\alpha)$ is a cone point of $\heartsuit W$. Such points are stabilized by involutions in $\boldsymbol{\Gamma}$. Hence α is also stabilized by an involution in $\boldsymbol{\Gamma}$. But then α cannot lie in the interior of $\boldsymbol{T}$. This is a contradiction. □

Now we know that the interior of $\heartsuit I$ lies in the interior of $\boldsymbol{T} \cap \boldsymbol{D}$ and that the far endpoint of $\heartsuit I$ lies on one of the two sides of $\boldsymbol{T}$ emanating from the origin.

LEMMA 19.5. *The endpoint of $\heartsuit I$ cannot lie on the diagonal edge of $\boldsymbol{T}$.*

Proof: Let b_2 be the endpoint of $\heartsuit I$. This point must lie on the diagonal edge of $\boldsymbol{T}$ which lies to the right of the piece $\clubsuit S_{14}$ from Figure 18.3, because $\heartsuit J$ does intersect the right edge of $\clubsuit S_{14}$ except at the endpoint $\mathbf{s}$. Figure 19.4 shows the situation. We check that all primary maps are defined and finite on a neighborhood of this segment. Hence h is defined in a neighborhood of b_2.

Since $\boldsymbol{R}(b_2) = b_2$, the point $h(b_2) \in \heartsuit W_1$ has a nontrivial stabilizer. This is only possible if $h(b_2)$ is the cone point of $\heartsuit W_1$. But then, by symmetry, we have

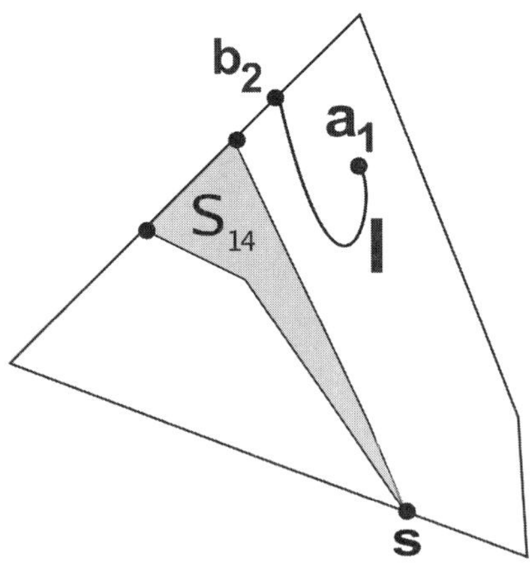

FIGURE 19.4. A_{14} blocks $\heartsuit I$ from reaching $\mathbf{t}$ for $\heartsuit X$

$h(\boldsymbol{R}(\heartsuit I)) \subset \heartsuit W_1$. But then, referring to Figure 19.3, the arc $\heartsuit I \cup \boldsymbol{R}(\heartsuit I)$ connects a_1 to a_4. But $h(a_4) = e$, the wrong vertex of $\boldsymbol{D}$. This this forces $\heartsuit W_1$ to connect $(1,1)$ to e rather than c. $\square$

Now we know that the endpoint of $\heartsuit I$ lies on the edge $\partial_2 \boldsymbol{T}$. Recall that $\partial_2 \boldsymbol{T}$ contains one tricky point, namely $\mathbf{s}$, where all the primary maps blow up. The point $\mathbf{s}$ was the last point we analyzed in the previous chapter. The point $\mathbf{s}$ is a vertex of $\clubsuit S_{14}$, so we know that the endpoint of $\heartsuit I$ lies somewhere on $\partial_2 \boldsymbol{T}$ to the right of $\mathbf{s}$. The next lemma is delicate, because there does turn out to be a (different) strand of $\heartsuit J$ connecting a_1 to the point $\mathbf{s}$.

LEMMA 19.6. $\mathbf{s}$ *is not the endpoint of* $\heartsuit I$.

Proof: The map h blows up $\mathbf{s}$ to the horizontal line λ_3 treated in §18.4. We claim that λ_3 does not intersect $\heartsuit W_1$. But the endpoint of $h(\heartsuit I)$ would be both in $\heartsuit W_1$ and in λ_3. This contradicts the claim.

To establish the claim, we observe that our analysis in §18.4 shows that $h(\mathbf{s})$ intersects $\heartsuit W$ at most 4 times. Recall from §18.4 that the orbit $\boldsymbol{\Gamma}(\lambda_3)$ intersects $\boldsymbol{T}$ in 4 arcs. The only arcs which intersect $\heartsuit W$ are the vertical segment and the second (arc of a) hyperbola. In both cases, there is only one intersection point, and this point lies in the interior of $\heartsuit W_1$ on the bottom strand of $\heartsuit J \cap \boldsymbol{T} - \boldsymbol{D}$. This gives potentially 4 intersection points of $\lambda_3 \cap \heartsuit W$ because 2 elements of $\boldsymbol{\Gamma}$ map λ_3 onto the line containing the vertical segment and two elements of $\boldsymbol{\Gamma}$ map λ_3 onto the second hyperbola. The 4 nontrival elements are

$$\boldsymbol{H}_{010}, \qquad \boldsymbol{H}_{100}, \qquad \boldsymbol{H}_{030}, \qquad \boldsymbol{H}_{120}.$$

Call these elements g_1, g_2, g_3, g_4. This analysis shows that

$$\lambda_3 \cap \heartsuit W \subset \bigcup_{i=1}^{4} g_i^{-1}(\heartsuit W_1^o). \tag{19.1}$$

None of the sets on the right hand side intersects $\heartsuit W_1$, as one can check by looking at the action of $\boldsymbol{\Gamma}$ on the vertices of $\heartsuit W$. This proves our claim. $\square$

Now we know that the endpoint of $\heartsuit I$ is some point $b_1 \in \partial_2 \boldsymbol{T}$ which lies to the right of $\mathbf{s}$. We use the area test to check that h is defined and a local diffeomorphism in a neighborhood of the open segment joining the right endpoint of $\partial_2 \boldsymbol{T}$ to $\mathbf{s}$. Hence h is a diffeomorphism in a neighborhood of b_1. In particular, $h(b_1)$ must be the cone point of $\heartsuit W_1$. By symmetry, $h(\heartsuit X) = \heartsuit W_1$, where

$$\heartsuit X = \heartsuit I \cup \gamma(\heartsuit I). \tag{19.2}$$

This is Property 1. By construction $\gamma(\heartsuit X) = \heartsuit X$. This is Property 2.

No point on the interior of $\heartsuit X$ is a point of $\heartsuit JC$. Hence no interior point of $\heartsuit X$ is a vertex of a $\boldsymbol{K}$-disk. We also know that $\heartsuit X$ can only enter a $\boldsymbol{K}$-disk at a vertex. Hence the interior of $\heartsuit X$ is disjoint from all the $\boldsymbol{K}$-disks. This establishes Properties 3-5. $\square$

19.2. From Generator to Edge

Now we move to Part 2 of the proof. In §12.5 we had 6 diffeomorphisms $h_i : \boldsymbol{K}_i \to \boldsymbol{D}$ for $i = 1, \ldots, 6$. Here we consider the semigroup $\langle h_1, h_2 \rangle$. With our current labeling scheme, we have $h_1 = \boldsymbol{H}_{101}$ and $h_2 = \boldsymbol{H}_{141}$. Recall that $\gamma = \boldsymbol{H}_{120}$ is the involution that preserves the generator $\heartsuit X$. We have $\gamma h_1 \gamma = h_2$.

Let $v_1 = (1,1)$ be the fixed point of h_1 and let v_2 be the fixed point of h_2. Let $\heartsuit JC_{12}$ denote the set of pre-images of $v_1 \cup v_2$ under the action of the semigroup $\langle h_1, h_2 \rangle$. There is a natural bijection between $\heartsuit JC_{12}$ and the set of infinite sequences with terms in $\{1, 2\}$. The endpoints of $\heartsuit X$ are the two sequences $122\ldots$ and $211\ldots$. The points in $\heartsuit JC_{12}$ form a Cantor subset of $\heartsuit JC$, and we identify this Cantor set with the middle third Cantor set, C_3. To describe the precise identification, we take point in $\heartsuit JC_{12}$ and change all the 1's to 0's. This gives us the base 3 expansion of the corresponding point in C_3. The endpoints of $\heartsuit X$ correspond to the points $1/3$ and $2/3$. Under this identification, h_1 is the dilation of C_3 which carries $C_3 \cap [0, 1/3]$ to C_3 and h_2 is the dilation which carries $C_3 \cap [2/3, 1]$ to C_3.

The map $\phi : \heartsuit JC_{12} \to C_3$ is a homeomorphism. Indeed, ϕ is Holder continuous, and so is the inverse. This follows from the metric expansion properties established in §12.7.

Let $\heartsuit PX$ denote the set of pre-images of $\heartsuit X$ under $\langle h_1, h_2 \rangle$. The elements of $\heartsuit PX$ are in bijection with the components of $[0, 1] - C_3$. For instance $\heartsuit X$ corresponds to the interval $[1/3, 2/3]$. Moreover, the elements of $\heartsuit PX$ are pairwise disjoint. For instance, $h_1^{-1}(\heartsuit X)$ is disjoint from $\heartsuit X$ because $h_1^{-1}(\heartsuit X)$ lies in the interior of $\boldsymbol{K}_1$ whereas $\heartsuit X$ is disjoint from the interior of $\boldsymbol{K}_1$. Figure 19.5 shows 3 elements of $\heartsuit PX$.

We pick some homeomorphism from $\heartsuit X$ to $[1/3, 2/3]$ and then extend ϕ so that $\phi(\heartsuit X) = [1/3, 2/3]$. We then extend ϕ to $\heartsuit PX \cup \heartsuit JC_{12}$ by requiring that

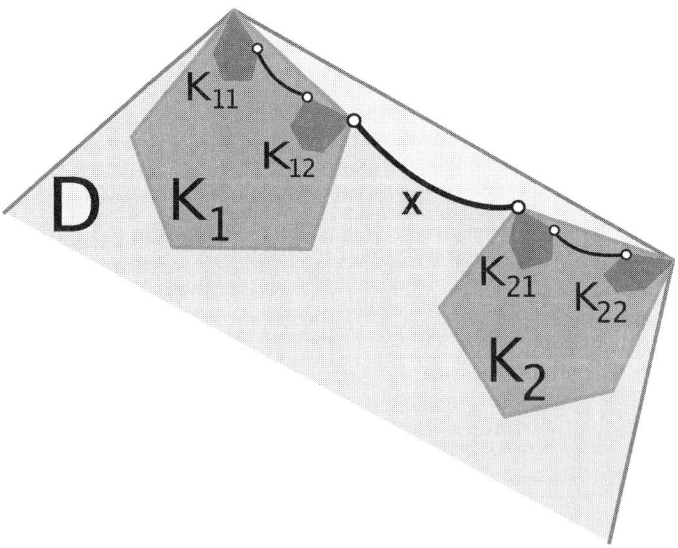

FIGURE 19.5. A few elements of $\heartsuit PX$

ϕ conjugate the action of $\langle h_1, h_2\rangle$ to the corresponding action on $[0,1]$. The image $\phi(\heartsuit PX)$ is the union of the closures of the components of $[0,1] - C_3$.

Finally, we let $\heartsuit Y$ denote the closure of $\heartsuit PX$. The continuity properties of ϕ guarantee that ϕ extends to all of $\heartsuit Y$. Since $\heartsuit Y$ is closed, $\phi(\heartsuit Y)$ is closed. Since $\phi(\heartsuit PX)$ is dense in $[0,1]$, we have that $\phi(\heartsuit Y) = [0,1]$. Hence $\phi : \heartsuit Y \to [0,1]$ is a continuous surjection. The fact that the components of $\heartsuit PX$ are pairwise disjoint guarantees that ϕ is also injective. Since $\heartsuit Y$ is compact and $[0,1]$ is Hausdorff, we see that ϕ is a homeomorphism. By construction $\gamma(\heartsuit Y) = \heartsuit Y$.

19.3. From Edge to Pentagon

We now consider the orbit $\boldsymbol{\Gamma}(\heartsuit Y)$. Recall that $\gamma \in \boldsymbol{\Gamma}$ is the element which stabilizes the side $\partial_2 \boldsymbol{T}$ of $\boldsymbol{T}$ and preserves $\heartsuit Y$. Since $\gamma(\heartsuit Y) = \heartsuit Y$, the orbit $\boldsymbol{\Gamma}(\heartsuit Y)$ consists of 5 embedded arcs. Each of these arcs connects 2 irregular fixed points of $\boldsymbol{H}$.

LEMMA 19.7. *The interior of $\heartsuit Y$ is contained in the interior of $\boldsymbol{T} \cup \gamma(\boldsymbol{T})$.*

Proof: Each gap in the middle third Cantor set is indexed by some finite binary sequence which denotes the composition of elements which maps this gap onto the middle third gap. The middle third gap, namely $[1/3, 2/3]$ is indexed by the empty string. For any finite string σ, the corresponding subset $\heartsuit X_\sigma$ of $\heartsuit PX$ is contained in the small disk $\boldsymbol{K}_\sigma$ obtained by pulling back $\boldsymbol{D}$ by the corresponding composition of elements. With this notation $\boldsymbol{K}_\emptyset = \boldsymbol{D}$ and $\boldsymbol{K}_1$ and $\boldsymbol{K}_2$ are as above. $\boldsymbol{K}_{11} = \boldsymbol{H}_1^{-1}(\boldsymbol{K}_1)$, and so on. Figure 19.5, which is a slightly enhanced version of Figure 19.6, shows the situation.

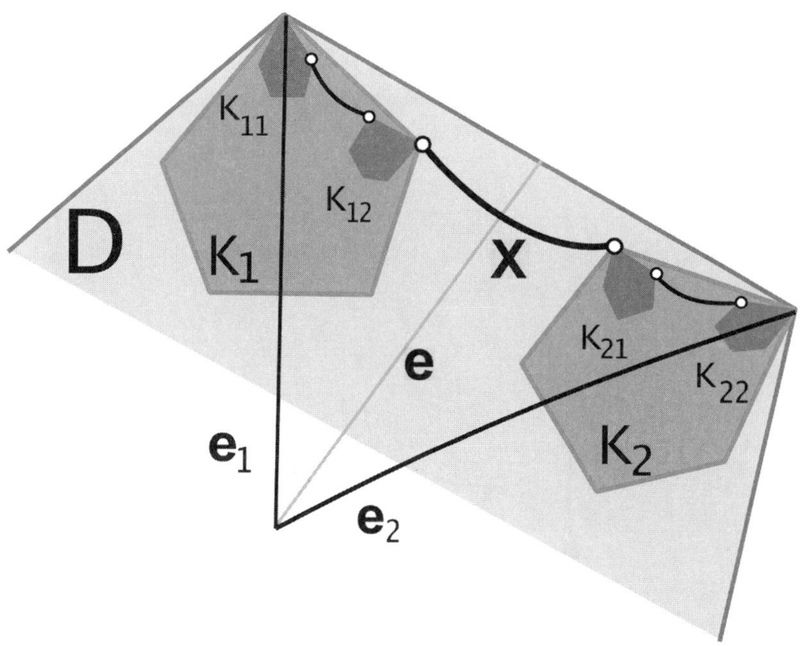

FIGURE 19.6. A few elements of $\heartsuit PX$

The diagonal edge of $\boldsymbol{T}$ is $\partial_0\boldsymbol{T}$. For this proof we set $e_1 = \partial_0\boldsymbol{T}$. The element h_1 preserves e_1. $\boldsymbol{T}$ and the element h_2 preserves the corresponding edge $e_2 = \gamma(e_1)$ of the adjacent fundamental domain $\gamma(\boldsymbol{T})$. These two arcs meet at the origin. By symmetry, e_j bisects $\boldsymbol{K}_\sigma$ exactly when σ consists entirely of js. Call such disks *bisected disks*. If σ does not consist entirely of a repeated digit, then $\boldsymbol{K}_\sigma$ is contained in the interior of $\boldsymbol{T} \cup \gamma(\boldsymbol{T})$. From this we see that all components of $\heartsuit PX$ lies in the interior of $\boldsymbol{T} \cup \gamma(\boldsymbol{T})$. The only points of $\heartsuit Y$ not in the interior of $\boldsymbol{T} \cup \gamma(\boldsymbol{T})$ are the ones contained in every bisected disk. But these are just the two endpoints. □

Given the way that $\boldsymbol{\Gamma}$ acts $\boldsymbol{T} \cup \boldsymbol{D}$, we see that $\boldsymbol{\Gamma}(\heartsuit Y)$ is an embedded pentagonal loop which connects the irregular fixed points of $\boldsymbol{H}$.

19.4. Pre-images of the Pentagon

Let L denote the disk[1] bounded by $\heartsuit \boldsymbol{\Gamma} Y$, and let $\heartsuit \partial L = \boldsymbol{\Gamma}(\heartsuit Y)$ denote its boundary. The pre-image $\boldsymbol{H}^{-1}(\heartsuit \partial L)$ consists of 6 embedded loops, $\heartsuit \partial L_1, \ldots, \heartsuit \partial L_6$.

Figure 19.7 illustrates the following lemma.

LEMMA 19.8. *The following is true for each $i = 1,2,3,4,5,6$. Let v_i be the vertex of $\heartsuit \partial L$ and $\heartsuit \partial L_i$. The two edges of $\heartsuit \partial L_i$ incident to v_i are contained in the two edges of $\heartsuit \partial L$ incident to v_i and the remaining points of $\heartsuit \partial L_i$ (except for the endpoints of the two special edges) are disjoint from $\heartsuit \partial L$.*

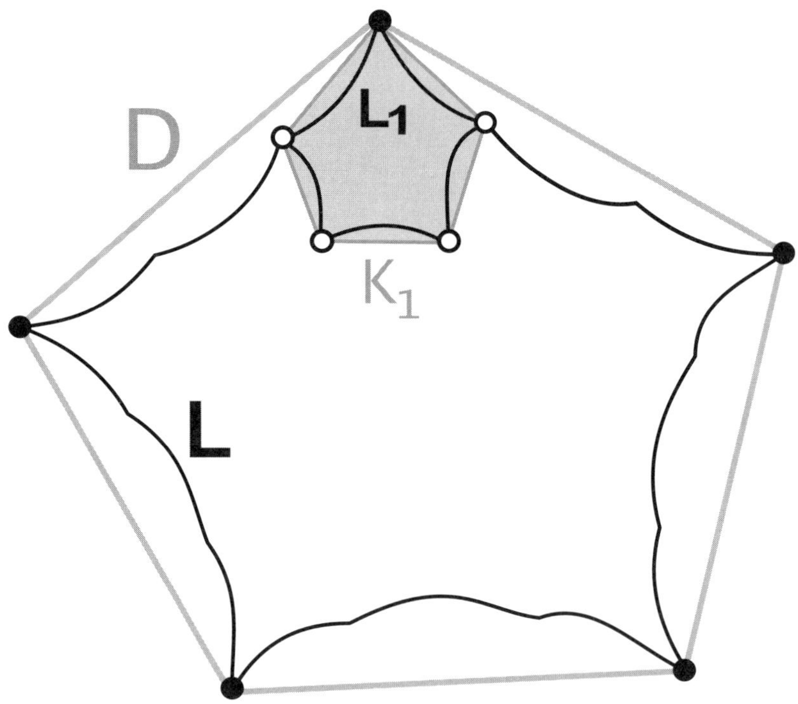

FIGURE 19.7. $\heartsuit \partial L$ and $\heartsuit \partial L_1$

[1]Our notation convention is such that we write L rather than $\heartsuit L$ because only ∂L is a subset of $\heartsuit J$. However, we will write $\heartsuit \partial L$.

Proof: We have

$$L_i \subset \boldsymbol{K}_i, \qquad i = 1, \ldots, 6. \tag{19.3}$$

Hence these loops are pairwise disjoint. Since $\boldsymbol{K}_3 \cap \heartsuit Y = \emptyset$ we have $\boldsymbol{K}_3 \cap \heartsuit\partial L = \emptyset$. Hence $\heartsuit\partial L_3 \cap \heartsuit\partial L = \emptyset$. Figure 19.7 shows $\heartsuit\partial L_1$ and $\heartsuit\partial L$.

By construction, $h_1(\heartsuit Y \cap \boldsymbol{K}_1) = \heartsuit Y$. But then $\heartsuit Y \cap \boldsymbol{K}_1$ is one of the edges of $\heartsuit\partial L_1$. A second edge of $\heartsuit\partial L_1$ is $\boldsymbol{R}(\heartsuit Y \cap \boldsymbol{K}_1)$. The remaining edges are disjoint from $\heartsuit\partial L$ because

$$\heartsuit\partial L - (\heartsuit Y \cap \boldsymbol{K}_1) - \boldsymbol{R}(\heartsuit Y \cap \boldsymbol{K}_1)$$

is disjoint from $\boldsymbol{K}_1$. □

Recall that $\heartsuit JC = \boldsymbol{B}(\mathcal{J}C)$, where $\mathcal{J}C$ is the Cantor set from Theorem 1.5.

LEMMA 19.9. *$\heartsuit JC \subset L$.*

Proof: The irregular fixed points of $\heartsuit\partial L$ are contained in L. They are the vertices of $\heartsuit\partial L$. Since $L_i \subset L$ for all i, we see by induction that any point of $\heartsuit JC$ in the grand pre-image of the irregular fixed points is contained in L. That is, $\heartsuit JC \cap L$ contains all the points which map to irregular fixed points under some power of $\boldsymbol{H}$. But this set is dense in $\heartsuit JC$ and L is closed. Hence $\heartsuit JC \subset L$. □

In the next chapter we will show the stronger result that $\heartsuit J \cap \boldsymbol{D} \subset L$.

19.5. The First Connector

Now we move to Part 3 of the proof. We find an arc of $\heartsuit J$ which remains in $\heartsuit J \cap \boldsymbol{D}$ and joins the point a_1 in Figure 19.8 to the cone point $\mathbf{s}$. We play the

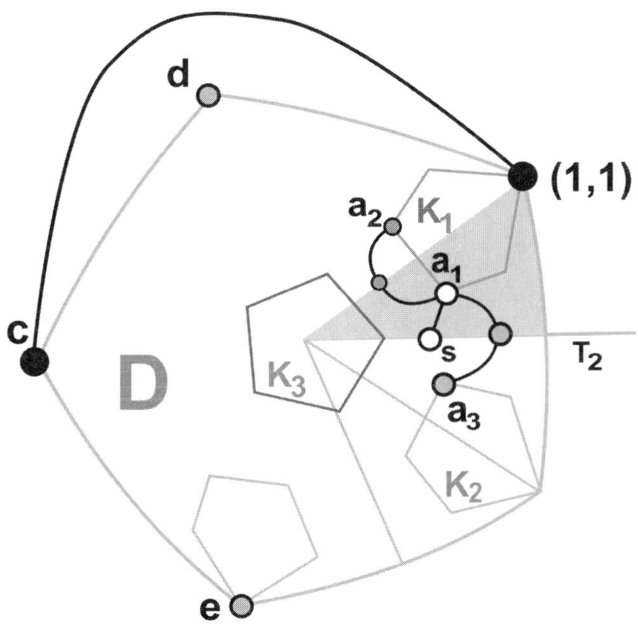

FIGURE 19.8. $\heartsuit\partial L$ and $\heartsuit\partial L_1$

same game that we played for the arc $\heartsuit X$, except that we change the names of some of the points and the elements. This time we let $h = \boldsymbol{H}_{011}$. We check that $h(a_1) = (1,1)$. We define $\heartsuit I$ as above.

The same arguments as above show that

- $\heartsuit I$ cannot exit $\boldsymbol{T} \cap \boldsymbol{D}$.
- An interior point of $\heartsuit I$ cannot lie on the boundary of $\boldsymbol{T}$.
- $\heartsuit I$ cannot have a cone point in the interior of $\boldsymbol{T} \cap \boldsymbol{D}$.
- $\heartsuit I$ cannot have its endpoint on the diagonal edge of $\boldsymbol{T}$. This time the problem is that $h(a_1) = d$ rather than c.
- $\heartsuit I$ cannot have its endpoint on a point of $\partial_2 \boldsymbol{T}$ to the left of $\mathbf{s}$ because the piece $\clubsuit S_{14}$ blocks it.
- $\heartsuit I$ cannot have its endpoint on a point of $\partial_2 \boldsymbol{T}$ that lies to the right of $\mathbf{s}$. This time the problem is that $h(a_3) = (1,1)$ rather than c.

The only possibility is that $\heartsuit I$ has its endpoint at $\mathbf{s}$. This is what we wanted. Since $\heartsuit I$ consists entirely of band points, except for its endpoints, we see that $\heartsuit I$ intersects the loops $\heartsuit \partial L$ and $\heartsuit \partial L_1, \ldots, \heartsuit \partial L_6$ only at the endpoint a_1.

19.6. The Second Connection

Now we find a connected arc that stays inside the partition piece $\clubsuit S_{13}$ and connects the two vertices $\mathbf{s}$ and $\mathbf{t}$. We repeat Figure 18.3 here and include a small side picture which shows all the previous arcs schematically.

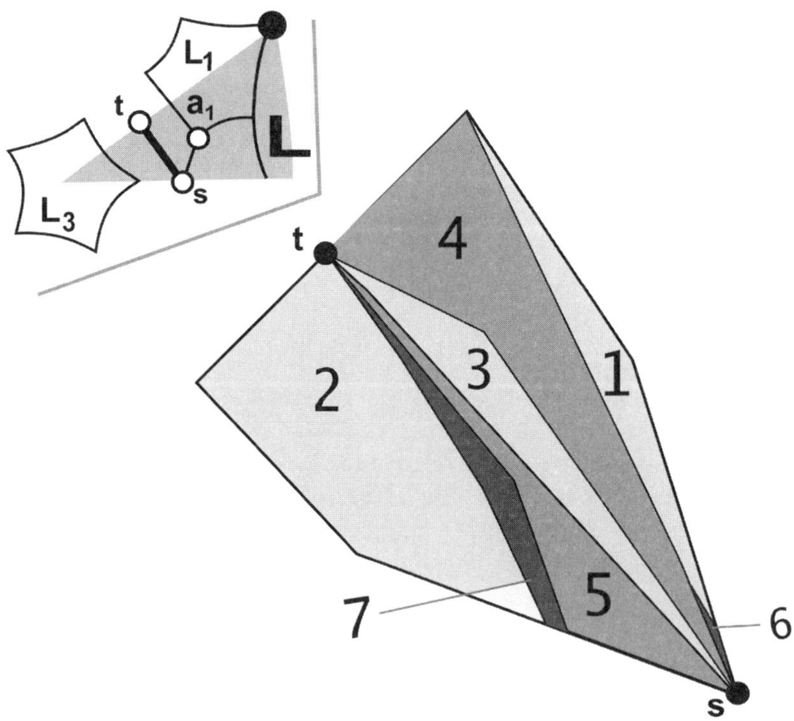

FIGURE 19.9. The pieces $\clubsuit S_{11}, \ldots, \clubsuit S_{17}$

LEMMA 19.10. *$\clubsuit S_{13} \cap \heartsuit J$ contains an arc which connects* $\mathbf{t}$ *to an interior point of $\clubsuit S_{13}$ and maps into the canonical loop under a primary element.*

Proof: The bottom strand β of $\heartsuit J \cap \boldsymbol{T} - \boldsymbol{D}$ is a subset of $\heartsuit W$. Let $h = \boldsymbol{H}_{041}$. This map blows $\mathbf{t}$ up to the vertical line ν through (ϕ^2, ϕ^2). Let e_1 and e_2 be the two lines through $\mathbf{t}$ containing the edges of $\clubsuit S_{13}$ incident to $\mathbf{t}$. Let $\mathbf{t}_1$ and $\mathbf{t}_2$ be the corresponding points on the exceptional fiber. We check that $h(\mathbf{t}_1)$ lies above the diagonal line, and hence above β. We also check that $h(\mathbf{t}_2)$ lies in the partition piece F, and hence below β. Finally, we check that $h(\mathbf{t}_3)$ lies between $h(\mathbf{t}_1)$ and $h(\mathbf{t}_2)$ for some point $\mathbf{t}_3$ on the exceptional fiber corresponding to a line that points into $\clubsuit S_{13}$.

Geometrically, this situation implies that h maps some initial segment e_1' of e_1 above β and h maps some initial segment e_2' of e_2 below β. (Here we mean that e_i' and e_2' both have $\mathbf{t}$ as an endpoint.) As long as we choose e_1' and e_2' small enough, h will map every segment connecting a point on e_1' to a point on e_2' to an arc which intersects β. Hence, points in $\clubsuit S_{13}$ arbitrarily close to $\mathbf{t}$ lie in the pre-image of $\heartsuit W$. But then, using the fact that $\mathbf{t}$ is a cone point, we see that these points are connected to $\mathbf{t}$ by arcs in $\heartsuit J$ once they are close enough to $\mathbf{t}$. □

Now we play the same game that we played for the first connector. We start with some small arc in $\clubsuit S_{13}$ that has one endpoint at $\mathbf{t}$ and lies in the pre-image of $\heartsuit W$. We let $\heartsuit I$ denote the maximal arc containing this initial arc and consisting entirely of band points.

LEMMA 19.11. *$\heartsuit I$ cannot exit $\clubsuit S_{13}$ except at the points* $\mathbf{s}$ *and* $\mathbf{t}$.

Proof: $\clubsuit S_{13}$ is flanked on either side by $\clubsuit S_{14}$ and $\clubsuit S_{15}$ which are disjoint from $\heartsuit J$ except at $\mathbf{s}$ and $\mathbf{t}$. This means that $\heartsuit I$ cannot exit $\clubsuit S_{13}$ except at $\mathbf{s}$ or $\mathbf{t}$. □

LEMMA 19.12. *$\heartsuit I$ cannot have an endpoint in the interior of $\boldsymbol{T} \cup \boldsymbol{D}$.*

Proof: This is the same argument as in Lemma 19.4. Such a point would have to be a fixed point of an involution in $\boldsymbol{\Gamma}$. □

LEMMA 19.13. *$\heartsuit I$ cannot make a loop connecting* $\mathbf{t}$ *to itself.*

Proof: The element $h_3 : \clubsuit S_{13} \to \boldsymbol{R}^2$ is a local diffeomorphism. The restriction $h_3 : \heartsuit I^o \to \boldsymbol{R}^2$ has the property that $h_3(\heartsuit I^o)$ consists entirely of band points. Hence $h_3(\heartsuit I^o)$ misses the points in $\boldsymbol{R}^2$ which we blew up to make the manifold $\boldsymbol{M}$. So, we can think of h_3 as a local diffeomorphism from $\heartsuit I^o$ to $\boldsymbol{M}$. But then $h_3(\heartsuit I^o)$ gives a regular parameterization of some arc of $\heartsuit W$. This arc connects two cone points of $\heartsuit W$ without hitting a vertex of $\heartsuit W$. This situation is impossible: Every two cone points of $\heartsuit W$ are separated by vertices of $\heartsuit W$. □

The only possibility is that $\heartsuit I$ connects $\mathbf{s}$ and $\mathbf{t}$. None of the previous arcs drawn can cross the edges of $\clubsuit S_{13}$, so our arc here is disjoint from all the previous arcs we have constructed, except that it shares the endpoint $\mathbf{s}$ with the arc from the previous section.

19.7. The Third Connector

There is a vertex a_2 of $\boldsymbol{K}_3$ on the edge $\partial_2\boldsymbol{T}$. In this section we find a connected arc of $\heartsuit J$ which joins a_2 to $\mathbf{t}$ and remains inside $\boldsymbol{T}\cap\boldsymbol{D}$. Figure 19.10 shows a schematic picture of this arc and the ones we have already described.

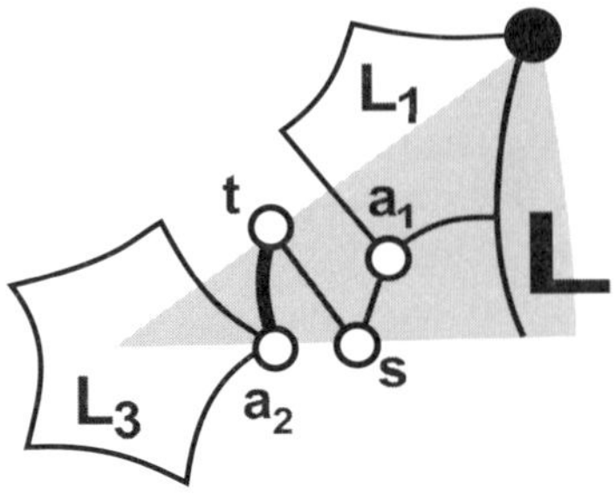

FIGURE 19.10. The three connectors

Let $h = \boldsymbol{H}_{111}$. We check that $h(a_2) = (1,1)$ and that h is a local diffeomorphism in a neighborhood of a_2. We also check that the component of $h^{-1}(\heartsuit W_1)$ ending on a_2 initially moves into $\boldsymbol{D}\cap\boldsymbol{T}$ from the boundary. We let $\heartsuit I$ denote the maximal strand that contains the initial part of this component and has an interior consisting entirely of band points. The same arguments as in previous sections show that the interior of $\heartsuit I$ lies in $\boldsymbol{T}\cap\boldsymbol{D}$ and the endpoint of $\heartsuit I$ lies on one of the two sides of $\boldsymbol{T}$.

LEMMA 19.14. *The endpoint of $\heartsuit I$ cannot lie on the edge $\partial_2\boldsymbol{T}$.*

Proof: We suppose that this is the case and derive a contradiction. Let α be the endpoint of $\heartsuit I$. The piece $\clubsuit S_{14}$ blocks $\heartsuit I$ from reaching and point on $\clubsuit S_{14}\cap\partial_2\boldsymbol{T}$. We check that h is defined and a local diffeomorphism in a neighborhood of the closed segment which connects $(0,0)$ to the left endpoint of $\clubsuit S_{14}$. But then h is defined at α and $h(\alpha)$ is a cone point.

Recall that γ is the involution which fixes $\partial_2\boldsymbol{T}$. The same arguments as above now say that h maps $\heartsuit I\cup\gamma(\heartsuit I)$ onto $\heartsuit W_1$, sending each endpoint of this doubled interval to an endpoint of $\heartsuit W_1$. But both endpoints of $\heartsuit I\cup\gamma(\heartsuit I)$ are a_2. This is a contradiction. □

Now we know that $\heartsuit I$ has its endpoint on the diagonal $\partial_0\boldsymbol{T}$.

LEMMA 19.15. *The endpoint of $\heartsuit I$ must be* $\mathbf{t}$.

Proof: Let Λ_0 and Λ_1 respectively denote the arcs of the main diagonal $\partial_0\boldsymbol{T}$ which join $\mathbf{t}$ to $(0,0)$ and $(1,1)$. The endpoint of $\heartsuit J$ certainly must lie on $\Lambda_1\cup\Lambda_2$, because $(0,0)$ lies in the interior of the small disk L_3 and $(1,1)$ is the far vertex of the small disk L_1. See Figure 19.10. We check that h is defined on the interior of Λ_1 and Λ_2 and h maps these two segment interiors into $\partial_2\boldsymbol{T}$, the bottom edge of $\boldsymbol{T}$. But every point of $\partial_2\boldsymbol{T}$ either lies in $\boldsymbol{D}$ or in a partition piece consisting entirely of preconvex points. Moreover, no point of $\partial_2\boldsymbol{T}$ is a vertex of $\boldsymbol{D}$. That is, $\partial_2\boldsymbol{T}$ does not intersect the orbit $\boldsymbol{\Gamma}(1,1)$. If the endpoint of $\heartsuit I$ lies in the interior of Λ_j, then h maps this endpoint into $\boldsymbol{D}-\boldsymbol{\Gamma}(1,1)$ and not into $\heartsuit W$. This is a contradiction. □

Since $\heartsuit I$ consists entirely of band points except for its endpoints, $\heartsuit I$ is disjoint from all the arcs we have already drawn, except at its endpoints.

Before we leave this section, we extract one more piece of information which will be useful later.

LEMMA 19.16. *The only point of $\heartsuit J$ on $\partial_0 \boldsymbol{T}$ strictly between the disks L_1 and L_3 is the point* $\mathbf{t}$.

Proof: We keep the notation from the previous lemma. Let Λ_0' denote the smaller arc of Λ_0 which connects a boundary point of L_3 to $\mathbf{t}$. Let Λ_1' denote the smaller arc of Λ_1 which connects a boundary point of L_1 to $\mathbf{t}$. We first consider Λ_0'. We use the cone test to check that the restriction of h to the interior of Λ_0 has the following property: $dh(1,1)\cdot(1,0) \geq 0$. So, the image $h(\Lambda_0')$ moves along $\partial_2 \boldsymbol{T}$ rightward from the image $h(\Lambda_0' \cap L_3)$, which is the rightmost point of $\partial_2 \cap \heartsuit J$. So, h maps the rest of the interior of Λ_0' to the right of this point, and no such point lies in $\heartsuit J$. The same argument works for Λ_1' except that this time we have $dh(1,1)\cdot(1,0) \leq 0$. $\square$

Remark: After reading the last proof, the reader might wonder how $\mathbf{t}$ manages to land in $\heartsuit J$. What is going on is that h blows up at $\mathbf{t}$, and arcs of $\heartsuit J$ reach $\mathbf{t}$ from directions not tangent to $\partial_0 \boldsymbol{T}$.

19.8. The End of the Proof

We let $\boldsymbol{Z}$ denote the union of the 3 arcs we have identified in the previous sections. Figure 19.11 shows $\boldsymbol{Z}$ as well as the loops $\heartsuit\partial L$ and $\heartsuit\partial L_1, \ldots, \heartsuit\partial L_6$. The union of these loops and $\boldsymbol{\Gamma}(\boldsymbol{Z})$ is precisely the seed for the graph $\mathcal{G}$. Note that $\boldsymbol{Z}$ is drawn with medium thick black lines in the figure.

The primary element h_i is invertible on the disk L_i and h_i^{-1} maps the seed we have constructed into a smaller copy of the seed whose outer loop is $\heartsuit\partial L_i$. We now

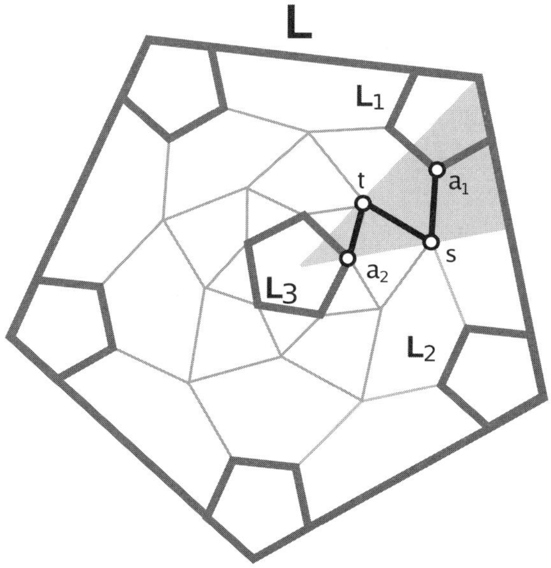

FIGURE 19.11. The seed

pull back these smaller seeds, and so on. Continuing this process indefinitely, we produce a homeomorphic copy $\heartsuit G$ of the graph $\mathcal{G}$.

By construction, all arcs of $\heartsuit G$ are obtained by pulling back arcs of the fundamental loop $\heartsuit W$. Moreover, $\heartsuit W$ is forward invariant in the sense that $h(p) \in \heartsuit W$ for each primary map h and each $p \in \heartsuit W$ where h is defined. Hence $\heartsuit G \cup \heartsuit W$ is forward invariant.

CHAPTER 20

Connectedness of the Julia Set

20.1. The Region Between the Disks

Let $\heartsuit G$ denote the embedded copy of the graph $\mathcal{G}$ considered in the last two chapters. We call the *main component* of $\heartsuit J$ the one which contains $\heartsuit G$. This component contains all of $\heartsuit G$ and also all of $\heartsuit J - \boldsymbol{D}$. Here we are using the fact that $\heartsuit J - \boldsymbol{D}$ is essentially a solenoid and hence connected. If $\heartsuit J$ is not path connected, then there is some second path component $\heartsuit \Psi$ and derive a contradiction.

Recall that L is the disk bounded by the pentagonal loop $\heartsuit \partial L$ constructed in the previous chapter, $\boldsymbol{D}$ is the big disk from §12.5, and $\heartsuit J$ is the Julia set. Our first step is to show that $\heartsuit \Psi$ cannot lie in $\boldsymbol{D} - L$. Indeed, we will prove the stronger result that $\heartsuit J$ is disjoint from

$$\boldsymbol{\Omega} = (\boldsymbol{D} - L) \cap \boldsymbol{T} \tag{20.1}$$

The set $\boldsymbol{\Omega}$ is a fundamental domain for the action of $\boldsymbol{\Gamma}$ on $\boldsymbol{D} - L$, so by symmetry $\heartsuit J$ cannot intersect $\boldsymbol{D} - L$.

We think of $\boldsymbol{\Omega}$ as a topological triangle, with vertices a, b, c.

(1) The side ac belongs to ∂D. We have already seen that ∂D intersects $\heartsuit J$ only at the vertices. Hence, $a = (1, 1)$ is the only point of ab which intersects $\heartsuit J$. Again, the proof is that the rest of $\boldsymbol{D} \cap \boldsymbol{T}$ is contained in partition pieces consisting entirely of preconvex points.

(2) The side bc belongs to the bottom edge $\partial_2 \boldsymbol{T}$ of the fundamental domain $\boldsymbol{T}$. Note that $\boldsymbol{H}_{111}(b) = \boldsymbol{q}$, the cone point on the main diagonal. This point is the lower vertex of $\clubsuit A$. Moreover, $\boldsymbol{H}_{111}$ maps the rest of $\boldsymbol{\Omega} \cap \partial_2 \boldsymbol{T}$ into the interior of the segment on the main diagonal bounded by the two special points $\boldsymbol{p}$ and $\boldsymbol{q}$. These points are contained in $\clubsuit A - \boldsymbol{q}$ and do not lie in $\heartsuit J$. Hence bc intersects $\heartsuit J$ only at b.

(3) The edge ab is half the arc $\heartsuit Y$ used to construct the disk L and its boundary $\heartsuit \partial L$. As the notation suggests ab lies entirely in $\heartsuit J$. The special point d in Figure 20.1 is the endpoint of the generator $\heartsuit X$. We will denote the edge ab by $\heartsuit \partial L \cap \boldsymbol{\Omega}$.

For the rest of this section, we set $h = \boldsymbol{H}_{141}$. In the section following this one, we prove the following result.

LEMMA 20.1 (Local Diffeomorphism). *The following is true.*

(1) *h is a local diffeomorphism from from the closure of $\boldsymbol{\Omega}$ into $\boldsymbol{R}^2$.*
(2) *$h(\boldsymbol{\Omega})$ is disjoint from the 3 points we blew up to create the manifold $\boldsymbol{M}$.*
(3) *$h(\boldsymbol{\Omega}) \cap L = \emptyset$.*

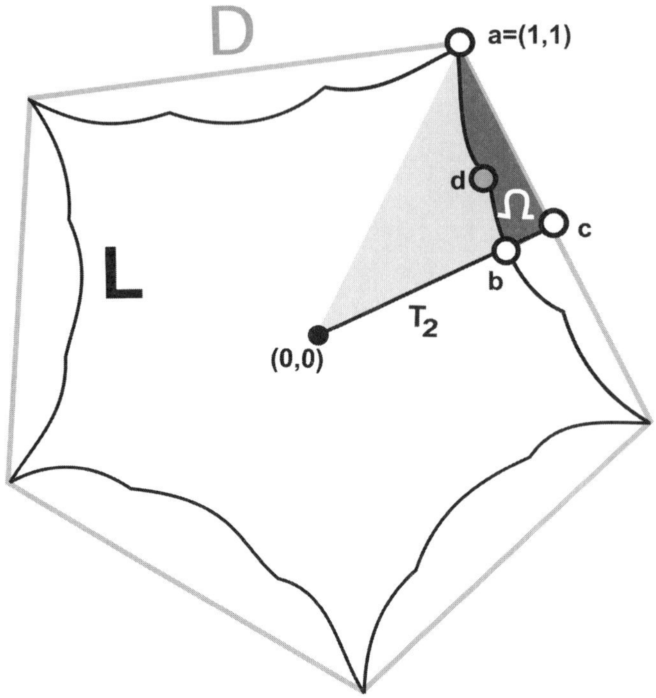

FIGURE 20.1. $\boldsymbol{\Omega}$ with its three vertices and special point highlighted

Definition: It might happen that $p \in \boldsymbol{\Omega}$ is such that $h(p) \in \boldsymbol{D} - L$. We call a point $p \in \boldsymbol{\Omega}$ *good* if $h(p) \in \boldsymbol{R}^2 - \boldsymbol{D}$.

LEMMA 20.2. *If $\heartsuit J \cap \boldsymbol{\Omega}$ contains a point, then $\heartsuit J \cap \boldsymbol{\Omega}$ contains a good point.*

Proof: Let $p_0 \in \heartsuit J \cap \boldsymbol{\Omega}$. Either $h(p_0) \in \boldsymbol{R}^2 - \boldsymbol{D}$ or else there is some $\gamma_0 \in \boldsymbol{\Gamma}$ such that $\gamma_0 h(p_0) = p_1 \in \heartsuit J \cap \boldsymbol{\Omega}$. Either $h(p_1) \in \boldsymbol{R}^2 - \boldsymbol{D}$ or else there is some $\gamma_1 \in \boldsymbol{\Gamma}$ such that $\gamma_1 h(p_1) = p_2 \in \heartsuit J \cap \boldsymbol{\Omega}$. Continuing this way we produce a sequence $p_0, p_1, p_2, \ldots$ Since $p_0 \notin \heartsuit JC$, Theorem 1.6 says that this sequence cannot go on forever. Eventually we have $p_n \in \heartsuit J\boldsymbol{A}$. So, some point in our sequence is good. $\square$

We suppose that $p \in \boldsymbol{\Omega} \cap \heartsuit J$ is good and we derive a contradiction. Since $\heartsuit JC \subset L$, we know that p is either a cone point or a band point. Just perturbing a little bit and using the openness of the goodness condition, we can assume without loss of generality that p is a band point.

LEMMA 20.3. *Some C^1 arc of $\heartsuit J \cap \boldsymbol{\Omega}$ connects p to a point of $\heartsuit \partial L \cap \boldsymbol{\Omega}$.*

Proof: Let $h = \boldsymbol{H}_{141}$. Each point of $\heartsuit J - \boldsymbol{D}$ is either a C^1 band point or a basic cone point. As we mentioned in §18.1, all the cone points here are $\boldsymbol{\Gamma}$ images of the ones covered by Lemma 17.1. By the Local Diffeomorphism Lemma, the same goes for all points of $\heartsuit J \cap \boldsymbol{\Omega}$ which are mapped by h into $\heartsuit J - \boldsymbol{D}$.

We know that $h(p) \in \heartsuit J - \boldsymbol{D}$. This means that all points of $\heartsuit J \cap \boldsymbol{\Omega}$ sufficiently close to p are C^1 band points or basic cone points. Suppose we choose the strand

$\heartsuit\beta$ of $\boldsymbol{J}$ through p and follow it around. First of all, note that $\heartsuit\beta$ cannot exit $\boldsymbol{\Omega}$ except by reaching L.

Suppose that $\heartsuit\beta$ never reaches L. Since $\heartsuit\beta$ cannot reach a point of the Cantor set $\heartsuit JC$, we see that $h(\heartsuit\beta)$ cannot reach a vertex of $\boldsymbol{D}$. The only pre-images of these vertices are in $\heartsuit JC$. So, $h(\heartsuit\beta)$ is stuck traveling around $\heartsuit J - \boldsymbol{D}$ in a C^1 way. But then $\heartsuit\beta$ endlessly travels around $\boldsymbol{\Omega}$ in a $\boldsymbol{C}^1$ way as well.

One possibility is that $\heartsuit\beta$ is a C^1 loop. But then $h(\heartsuit\beta)$ is a C^1 loop of $\heartsuit J - \boldsymbol{D}$. There are no such loops, essentially by Theorem 1.10. Any C^1 path in $\heartsuit J - \boldsymbol{D}$ lifts to a path in the solenoid, and the solenoid has no closed loops. So, this situation is impossible.

The other possibility is that $\heartsuit\beta$ is a bi-infinite C^1 path. But then $h(\heartsuit\beta)$ is a a bi-infinite C^1 path in $\heartsuit J - \boldsymbol{D}$ which never exits $\boldsymbol{R}^2$. This contradicts Lemma 17.5.

This contradiction shows that $\heartsuit\beta$ must reach L. □

Remark: This last result, on its own, rules out an extra connected component in $\boldsymbol{D} - \boldsymbol{L}$. We do not stop with this result because we are also interested in knowing, for its own sake, that $\heartsuit J$ does not intersect $\boldsymbol{\Omega}$. The reader who only cares about the connectedness of $\heartsuit J$ can skip the rest of this section.

Continuing with the notation from the previous proof, we let $\heartsuit\beta$ be a C^1 path in $\heartsuit J \cap \boldsymbol{\Omega}$ which connects a point $x \in \heartsuit\partial L \cap \boldsymbol{\Omega}$ to the good point p. Now we inquire more carefully about the nature of the point x. We trim $\heartsuit\beta$ to that the interior of $\heartsuit\beta$ is contained in the interior of $\boldsymbol{\Omega}$.

Figure 20.2 shows the kind of conclusion we are trying to draw about β. The 3 arcs in Figure 20.2 emanating from p represent possibilities for β. Figure 20.2 also shows an auxiliary arc $\heartsuit Z$ which will play an important role in the next lemma.

LEMMA 20.4. *x is one of the two vertices of $\boldsymbol{\Omega}$ which lie on L.*

Proof: Note that x cannot lie in the interior of $\heartsuit X$, because the interior of this arc consists entirely band points. Nothing can attach to it. So, if x is not one of the vertices of $\boldsymbol{\Omega}$, or the special point, then $x \in \boldsymbol{K}_1$. (Here $\boldsymbol{K}_1$ is one of the 6

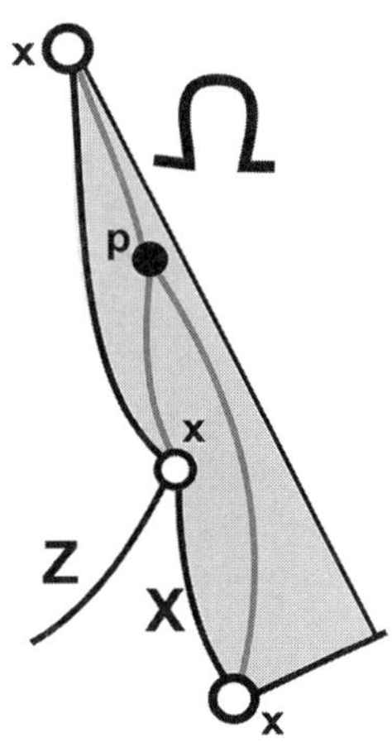

FIGURE 20.2. β connecting p to one of 3 endpoints

pre-images of $\boldsymbol{D}$ under primary maps constructed in §12.5 and considered in the last chapter. See Figure 19.6.) In other words, referring to the homeomorphism ϕ from $\heartsuit Y$ to $[0,1]$ constructed in the last chapter, $\phi(x) < 1/3$. This gives us the following situation

$$h(x) \in \boldsymbol{D}^o, \qquad h(p) \in \boldsymbol{R}^2 - \boldsymbol{D}. \tag{20.2}$$

This means that $h(\heartsuit\beta)$ connects a point in the interior of $\boldsymbol{D}$ to a point outside $\boldsymbol{D}$. But then $h(\heartsuit\beta)$ contains a vertex of $\boldsymbol{D}$ in its interior. But then β contains a point of $\heartsuit JC$ in its interior. This is a contradiction.

We have not yet ruled out the possibility that x is the special point of $\boldsymbol{\Omega}$. We will assume this and derive a contradiction. The property we will use is that there are only two arcs of $\heartsuit J \cap \boldsymbol{T} - \boldsymbol{D}$ which limit on $(1,1)$. One of these arcs is the bottom strand of $\heartsuit J \cap \boldsymbol{T} - \boldsymbol{D}$ and the other one is the reflection of this strand in the main diagonal.

Call an arc $\heartsuit A$ *special* if $\heartsuit A$ limits on x and $\heartsuit \boldsymbol{H}_{031}(A) \in \heartsuit J \cap \boldsymbol{T} - \boldsymbol{D}$. There are only 2 special arcs because $\boldsymbol{H}_{031}$ is a diffeomorphism in a neighborhood of x which maps x to $(1,1)$. One of the special arcs is $\heartsuit X$. We check that $\boldsymbol{H}_{031}$ is an orientation reversing diffeomorphism near x, and this forces the second special arc, $\heartsuit Z$, to lie outside $\boldsymbol{\Omega}$. But our arc $\heartsuit\beta$ is also special, and this would give a third special arc. This is a contradiction. □

Starting at p, we can follow the arc β in either direction. Applying the above lemma to both endpoints, we see that the arc β contains the good point $p \in \heartsuit J \cap \boldsymbol{\Omega}$ and has both endpoints at the vertices of $\boldsymbol{\Omega}$.

LEMMA 20.5. *At least one endpoint of $\heartsuit\beta$ must be $(1,1)$.*

Proof: If this is false, then both endpoints of $\heartsuit\beta$ lie on the cone point

$$x = \partial_2 \boldsymbol{T} \cap \heartsuit \partial L.$$

Since h is a local diffeomorphism even at the endpoints of $\heartsuit\beta$, we can extend $\heartsuit\beta$ to a slightly longer path $\widehat{\heartsuit}\beta$ and consider the image $h(\widehat{\heartsuit}\beta)$. If we just extend a little bit we can guarantee that $h(\widehat{\heartsuit}\beta) \subset \boldsymbol{R}^2$. The path $h(\widehat{\heartsuit}\beta)$ contains some points of $\heartsuit J - \boldsymbol{D}$ and no vertices of $\boldsymbol{D}$. Hence $h(\widehat{\heartsuit}\beta) \subset \heartsuit J - \boldsymbol{D}$. This path is a regular parameterization of some C^1 arc of $\heartsuit J \cap \boldsymbol{D}$ which continues around long enough to encounter the same cone point twice. But then $h(\widehat{\heartsuit}\beta)$ must enter into at least 11 consecutive tiles of $\boldsymbol{\Gamma}(\boldsymbol{T})$. By Lemma 17.5 the path $h(\widehat{\heartsuit}\beta)$ contains some points not in $\boldsymbol{R}^2$. This is a contradiction. □

Now we come to the final contradiction. We have

$$\boldsymbol{H}_{141}(\beta^o) \subset \heartsuit J - \boldsymbol{D}. \tag{20.3}$$

Here β^o is the set of points of β other than the endpoints.

But consider a very short initial arc $\heartsuit\beta'$ of $\heartsuit\beta$ which emanates from $(1,1)$. Since $h = \boldsymbol{H}_{141}$ is a local diffeomorphism in a neighborhood of $(1,1)$, so is every other primary element. In particular, this is true of the primary element $\boldsymbol{H}_{101}$. What does this element do to $\heartsuit\beta'$? Note that $\boldsymbol{H}_{101}$ fixes $(1,1)$ and maps a neighborhood of $\heartsuit J - \boldsymbol{D}$ diffeomorphically over itself near $(1,1)$. More precisely, near $(1,1)$ the set $\heartsuit J \cap \boldsymbol{D}$ lies between the bottom strand of $\heartsuit J$ emanating from $(1,1)$ and to the right of the leftmost strand of $\heartsuit J$ emanating from $(1,1)$. The element

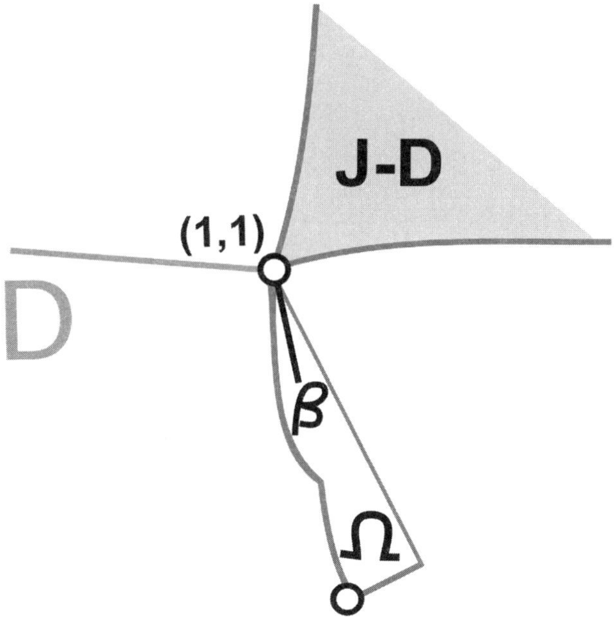

FIGURE 20.3. ∂L and ∂L_1

$\boldsymbol{H}_{101}$ maps these two extremal strands into themselves. Figure 20.3 shows the situation.

But then

$$\boldsymbol{H}_{101}(\heartsuit\beta') \not\subset \heartsuit J \cap \boldsymbol{D}.$$

But then, by symmetry, the same goes for $\boldsymbol{H}_{141}(\heartsuit\beta')$. This contradicts equation (20.3).

20.2. The Local Diffeomorphism Lemma

Now we prove the Local Diffeomorphism Lemma.

Let P be the convex hull of $\clubsuit S_2 \cup \mathbf{s}$. In Figure 20.4 below, P is the darkly shaded pentagon. Our idea is to show that P has a special arc B running through it that serves as a barrier between $\boldsymbol{\Omega}$ and the tricky piece $\clubsuit S_1$. Here we study how the map $h = \boldsymbol{H}_{131}$ acts on $P' = P' - \mathbf{s}$.

LEMMA 20.6. *h is defined, finite, and a local diffeomorphism on P. Moreover, throughout $P - \mathbf{s}$ the differential maps lines of slope in $[0, 1/2]$ to timelike lines.*

Proof: Let $h = \boldsymbol{H}_{131}$. We use the denominator test coupled with the subtraction trick to show that h is defined and finite on $P - \mathbf{s}$. We use the area test to check that that $\det(dh) \geq \phi^{-2}$ on P'. Hence h is a local diffeomorphism on P'. We use the cone test to check the last claim. □

We define the *top edge* of P to be the edge of P opposite $\mathbf{s}$ in the combinatorial sense. This edge has slope in $(0, 1/2)$.

LEMMA 20.7. *$h^{-1}(\partial_2 \boldsymbol{T}) \cap P$ is a smooth arc connecting $\mathbf{s}$ to the top edge of P.*

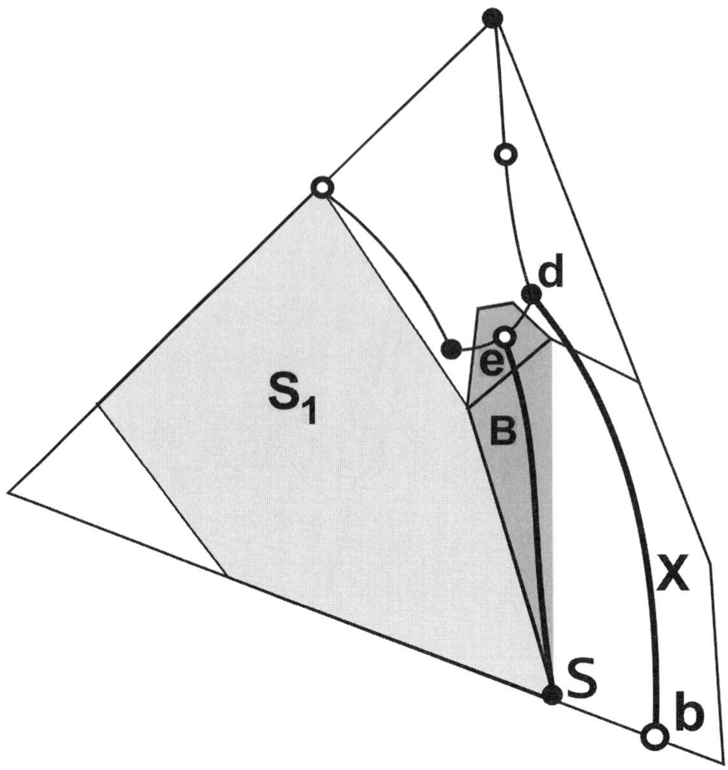

FIGURE 20.4. The pentagon P and the curve B

Proof: We define the *left edges* of P to be the two edges connecting $\mathbf{s}$ to the top edge and staying to the left. We define the *right edges* of P similarly.

We use the exclusion test to check that h maps the left edges of P entirely below $\partial_2 \boldsymbol{T}$ and the right edges of P entirely above $\partial_2 \boldsymbol{T}$. We use the confinement test to check that

$$h(P') \subset [0, \phi^2] \times [-20, 20].$$

Hence $h(P')$ is contained in strip bounded by the lines $x = 0$ and $x = \phi^6$, which is the vertical edge of $\boldsymbol{T}$. These properties, together with the timelike nature of the images of the leaves of our foliation, guarantee that each leaf intersects $h^{-1}(\partial_2 \boldsymbol{T})$ exactly once. Hence this set is a curve which connects $\mathbf{s}$ to the top edge of P. Since h is a local diffeomorphism, this curve is smooth. □

LEMMA 20.8. *$h^{-1}(\partial_2 \boldsymbol{T})$ contains an arc B such that $h(B)$ has one endpoint at the vertex b of $\boldsymbol{\Omega}$ and otherwise lies to the right of b on $\partial_2 \boldsymbol{T}$.*

Proof: Since h is a local diffeomorphism in P' we see that the restriction of h to $h^{-1}(\partial_2 \boldsymbol{T})$ gives a regular parametrization of a segment of $\partial_2 \boldsymbol{T}$. In other words, h is injective on $h^{-1}(\partial_2 \boldsymbol{T})$. We check by direct calculation that h maps some point $e \in ♣S_2$ to b. Let B be the portion of $h^{-1}(\partial_2 \boldsymbol{T})$ connecting $\mathbf{s}$ to e. Given the injectivity of h we see that $h(B - e)$ either pies to the left of b or to the right. Given that h is orientation preserving in P' we see that $h(B - e)$ lies to the right of b. □

LEMMA 20.9. *The arc B intersects $\heartsuit J$ only at its endpoints.*

Proof: It suffices to prove that there are no points of $\boldsymbol{J}$ on the bottom edge $\partial_2\boldsymbol{T}$ which lie to the right of b. We already know that the side of bc of $\boldsymbol{\Omega}$ contains no points of $\heartsuit J$ except b. The points of $\partial_2\boldsymbol{T}$ to the right of $\partial\boldsymbol{D}$ lie in $\clubsuit B \cup \clubsuit F \cup \spadesuit C$. We have already seen that this union is disjoint from $\heartsuit J$. □

There are 5 special cone points on $\heartsuit\partial L$ which are fixed points of involutions in $\boldsymbol{\Gamma}$. These points come between the vertices of $\heartsuit\partial L$. We call these points the *symmetric cone points*. We define 5 symmetric cone points on each loop $\heartsuit\partial L_i$ by pulling back the symmetric cone points by suitable primary elements.

LEMMA 20.10. *$\heartsuit\partial L \cap \boldsymbol{\Omega}$ is disjoint from $\clubsuit S_1$.*

Proof: Our argument refers to Figure 20.4. We have

$$\heartsuit\partial L \cap \boldsymbol{\Omega} \subset \heartsuit X \cup \heartsuit\partial L_1. \tag{20.4}$$

Since $L_1 \subset \boldsymbol{K}_1$ and $\boldsymbol{K}_1 \cap \clubsuit S_1 = \emptyset$ the last set in equation (20.4) is disjoint from $\clubsuit S_1$. Hence, it suffices to prove that $\heartsuit X \cap \clubsuit S_1 = \emptyset$. This is what we will do.

Consider the arc B. The endpoint e of B is a symmetric cone point of $\heartsuit\partial L_1$ because $h(e) = b$, which a symmetric cone point of $\heartsuit L$. The generator $\heartsuit X$ starts to the right of B at the vertex d of L_1 and ends to the right of B at a point $c \in \partial_2\boldsymbol{T}$. Given that $\heartsuit X$ is disjoint from L_1 except at its endpoint, we see that $\heartsuit X$ must stay to the right of B. But then B blocks $\heartsuit X$ from entering and partition pieces to the right of B. This includes the partition piece $\clubsuit S_1$. □

Let

$$\clubsuit S' = \clubsuit S_2 \cup \clubsuit S_3 \cup \clubsuit T \tag{20.5}$$

denote the union of all the tiles of $\clubsuit Q$ which lie to the right of $\clubsuit S_1$. By the result in the previous section, $\boldsymbol{\Omega} \subset \clubsuit S' - \mathbf{s}$. We use the area test to show that $h = \boldsymbol{H}_{141}$ is a local diffeomorphism on $\clubsuit S' - \mathbf{s}$. Using the confinement test, we check that $h(\clubsuit S' - \mathbf{s})$ lies beneath the line λ_3 studied in §18.4 and to the left of the y-axis. This quadrant contains no blow-up points.

To finish the proof of the Local Diffeomorphism Lemma we just have to show that $h(\boldsymbol{\Omega})$ lies outside the disk L. We already have shown that all primary elements map the pieces $\clubsuit S_2$ and $\clubsuit S_3$ outside the larger disk $\boldsymbol{D}$. So, we just have to show that $h(\boldsymbol{\Omega} \cap \clubsuit T)$ lies outside L.

In the proof of Theorem 1.5, and we showed that $\boldsymbol{H}_{101} : \clubsuit T \to \boldsymbol{R}^2$ is a diffeomorphism. Since $\boldsymbol{H}_{101}(L_1) = L$ and $\boldsymbol{\Omega}$ lies outside of L_1, we see that

$$\boldsymbol{H}_{101}(\boldsymbol{\Omega} \cap \clubsuit T) \cap L = \emptyset. \tag{20.6}$$

Since L is $\boldsymbol{\Gamma}$ invariant, we see that this same result holds with h in place of $\boldsymbol{H}_{101}$. This finishes the proof of the Local Diffeomorphism Lemma.

20.3. A Case by Case Analysis

Now we know that our bad component Ψ lies somewhere in the disk L. Using the action of $\boldsymbol{H}$ we can assume without loss of generality that $\heartsuit\Psi \subset L - \bigcup L_i$. That is, $\heartsuit\Psi$ is entirely contained in the region we called the *seed* at the end of §19. See Figure 19.1119.10.

Note that $\heartsuit\Psi$ must be contained in the complement of $\heartsuit G$. Looking at Figure 19.11 we see that $\heartsuit\Psi$ must be contained in two consecutive fundamental domains for $\boldsymbol{\Gamma}$. Moreover, by Lemma 19.16, the only point in the seed which lies on the main diagonal and in $\heartsuit J$ is the point $\mathbf{t}$. Hence $\heartsuit\Psi$ cannot intersect the diagonal edge of $\boldsymbol{T}$. From this fact, and from symmetry, we can assume that

$$\heartsuit\Psi \subset \boldsymbol{T} \cup \gamma(\boldsymbol{T}), \tag{20.7}$$

where $\gamma \in \boldsymbol{\Gamma}$ is the stabilizer of the $\partial_2\boldsymbol{T}$

Now we rule out the various possibilities for where the bad component $\heartsuit\Psi$ might be. The pieces here refer to the partition of the piece $\clubsuit S_1$ used in the proof of Theorem 1.8. See Figure 19.9 or Figure 20.5 below. We treat these pieces in an order which makes the proofs run as smoothly as possible. Lemma 17.5 is going to be the workhorse for us. Again, this result says that an infinite C^1 path in $\heartsuit J - \boldsymbol{D}$ must exit $\boldsymbol{R}^2$. Moreover $\heartsuit J - \boldsymbol{D}$ has no C^1 loops.

LEMMA 20.11. *$\heartsuit\Psi$ cannot intersect $\clubsuit S_{14}$, or $\clubsuit S_{15}$.*

Proof: We have seen already that $\heartsuit J$ only intersects $\clubsuit S_{14}$ and $\clubsuit S_{15}$ at the points $\mathbf{s}$ and $\mathbf{t}$ which belong to the main component. □

In order to apply Lemma 17.5 effectively, we want to avoid the points of $\boldsymbol{R}^2$ which we blew up to get the manifold $\boldsymbol{M}$. We call a map $h : X \to \boldsymbol{R}^2$ *clean* if $h(X)$ does not contain the 3 points we blew up to create the manifold $\boldsymbol{M}$.

LEMMA 20.12. *$\heartsuit\Psi$ cannot intersect $\clubsuit S_{13}$.*

Proof: The only points on $\partial\clubsuit S_{13}$ which do not lie in $\clubsuit S_{14} \cup \clubsuit S_{15}$ are the vertices $\mathbf{s}$ and $\mathbf{t}$. So, if $\heartsuit\Psi$ intersects $\clubsuit S_{13}$ then $\heartsuit\Psi \subset \clubsuit S_{13}^o$. Our proof in Lemma 18.4 shows that every point of $\heartsuit J \cap \clubsuit S_{13}^o$ is either a C^1 band point or a basic cone point. That is, the strands of $\heartsuit J$ can be continued through the cone points uniquely in a C^1 way.

Let $h_3 = \boldsymbol{H}_{131}$. We showed already that h_3 is a local diffeomorphism on $\clubsuit S_{13}$, and in the remark following Lemma 18.4 we checked that h_3 is clean on $\clubsuit S_{13}^o$. Finally, we know that $h_3(\clubsuit S_{13})$ is disjoint from $\boldsymbol{D}$.

Now let $\heartsuit\beta$ be a maximal C^1 arc in $\heartsuit J \cap \heartsuit\Psi$. We get $\heartsuit\beta$ by moving along the strands of a Cantor band until we reach a cone point, then continuing through in a C^1 way, and so on. If $\heartsuit\beta$ exits $\clubsuit S_{13}$ then $\heartsuit\beta$ contains $\mathbf{s}$ or $\mathbf{t}$ and we see that $\heartsuit\Psi$ is part of the main component. So, $\heartsuit\beta$ must either be a loop or a bi-infinite path contained in the interior of $\clubsuit S_{13}$. We will rule out either case.

Since h_3 is a local diffeomorphism that is clean on $\heartsuit\beta$, the image $h_3(\heartsuit\beta)$ is a regular parameterization of a C^1 arc of $\heartsuit J - \boldsymbol{D}$. Since there are no C^1 loops in $\heartsuit J - \boldsymbol{D}$ we see that $\heartsuit\beta$ cannot be a loop. Hence $h_3(\heartsuit\beta)$ is a bi-infinite path. But then $h^3(\heartsuit\beta)$ leaves $\boldsymbol{R}^2$ by Lemma 17.5. This contradicts the fact that $h_3(\heartsuit\beta) \subset \boldsymbol{R}^2$. □

LEMMA 20.13. *$\heartsuit\Psi$ cannot intersect $\clubsuit S_{17}$.*

Proof: We know from Lemma 18.9 that $\heartsuit J$ does not intersect $\partial\clubsuit S_{17} - \mathbf{t}$. Hence $\heartsuit\Psi$ cannot intersect $\partial\clubsuit S_{17}$. If $\heartsuit\Psi$ intersects $\clubsuit S_{17}$, then $\heartsuit\Psi \subset \clubsuit S_{17}^o$. The analysis in §18.3 shows that every point of $\heartsuit J \cap \clubsuit S_{17}$ is either a C^1 band point or a basic cone point. That is, the tangent lines are defined even at the cone points. Thus, we may choose a maximal C^1 arc of $\heartsuit\Psi$. From the analysis in §18.3 we see that this arc is transverse to the horizontal foliation. Hence, there is one direction of or arc which must exit at $\mathbf{t}$. But then $\heartsuit\Psi$ is in the same path component as the main path component. □

Let $\clubsuit X$ denote the region of the piece $\clubsuit S$ which is the convex hull of the origin and $\clubsuit S_{12}$. In other words, $\clubsuit X$ is everything in $\clubsuit S$ lying to the left of the right edges of $\clubsuit S_{12}$. Put another way, $\clubsuit X$ is the union of $\clubsuit S_{12}$ and everything to the left of $\clubsuit S_1$. The shaded region in Figure 20.5 is $\clubsuit X$.

LEMMA 20.14. *$\heartsuit\Psi$ cannot intersect $\clubsuit X$.*

Proof: We first outline the proof and then we give the details.

(1) We show that $\heartsuit\Psi$ is contained in the interior of $\clubsuit Y = \clubsuit X \cup \gamma(\clubsuit X)$.
(2) Let $h = \boldsymbol{H}_{111}$. We show that h is a clean local diffeomorphism on $\clubsuit Y^o$.
(3) We show that $h(\heartsuit\Psi) \cap \boldsymbol{D} = \emptyset$.
(4) We use the same C^1 arc trick as in the previous proofs.

Step 1: The right edges of $\clubsuit S_{12}$ coincide with the left edges of $\clubsuit S_{17}$, and we have shown that the left edges of $\clubsuit S_{17}$ do not intersect $\heartsuit J$ except at the point $\mathbf{t}$. So, if $\heartsuit\Psi$ intersects $\clubsuit X$, then $\heartsuit\Psi \subset \clubsuit Y$. Now, the boundary of $\clubsuit Y$ consists of points on the diagonal, points on the image of the diagonal under γ, and points on the right edge of $\clubsuit S_{12}$. From what we have already said above, and symmetry, $\heartsuit\Psi$ cannot intersect any of these sets. Hence $\heartsuit\Psi \subset \clubsuit Y^o$.

Step 2: We use the area test to show that $-\det(dh) \geq \phi^{-1}$ on $\clubsuit X - \mathbf{t}$. Hence h is a local diffeomorphism on $\clubsuit X - \mathbf{t}$. We use the confinement test to check that

$$h(\clubsuit X - \mathbf{t}) \subset \boldsymbol{T} - \clubsuit A, \tag{20.8}$$

This piece contains no blow-up points. Hence h is a clean map. From the equation $h \circ \gamma = \boldsymbol{R} \circ h$ and the fact that $\boldsymbol{R}$ preserves the set of blow-up points, we see by symmetry that h is a clean local diffeomorphism on $\clubsuit Y^o$.

Step 3: We have already seen in §12.3 that h is a diffeomorphism on the portion of $\clubsuit X$ lying to the left of $\clubsuit S_{12}$. This is the partition piece $\clubsuit R$. Moreover, h maps the disk L_3 onto L. Since $\heartsuit\Psi$ lies outside L_3, we conclude that $h(\heartsuit\Psi \cap \clubsuit X - \clubsuit S_{12})$ lies outside L. Since this set lies in $\heartsuit J$, we see that

$$h(\heartsuit\Psi \cap \clubsuit X - \clubsuit S_{12}) \cap \boldsymbol{D} = \emptyset.$$

If $h(\heartsuit\Psi) \cap \boldsymbol{D}$ contained a point, then we could connect a point in $\heartsuit J \cap \boldsymbol{D}$ to a point in $\heartsuit J - \boldsymbol{D}$. by a path in $h(\heartsuit\Psi)$. But then some point of $h(\heartsuit\Psi)$ would be a vertex of $\boldsymbol{D}$. But then some point of $\heartsuit\Psi$ would lie in $\heartsuit \boldsymbol{JC}$. This is a contradiction. Hence $h(\heartsuit\Psi \cap \clubsuit X) \cap \boldsymbol{D} = \emptyset$. By symmetry, the same goes for $h(\heartsuit\Psi \cap \gamma(\clubsuit X))$.

Step 4: Now we have the usual contradiction. All the points of $\heartsuit\Psi$ are C^1 band points and basic cone points. So we let $\heartsuit\beta$ be a maximal C^1 arc in $\heartsuit\Psi$. This arc must either be a loop or a bi-infinite path. Either case leads to the same contradiction as in previous lemmas. □

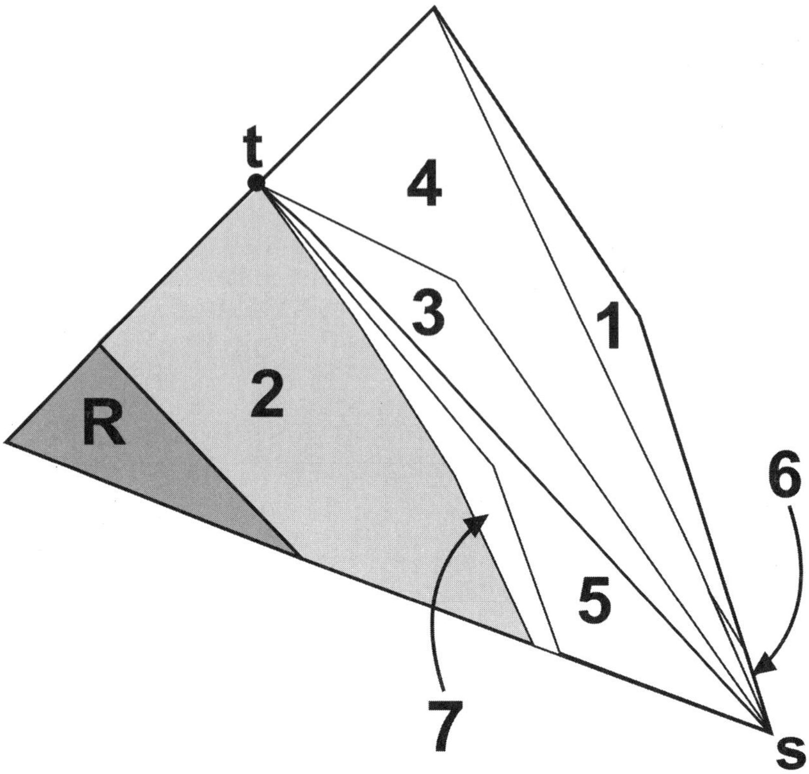

FIGURE 20.5. The set $\clubsuit X$ is shaded. The numerical label k denotes $\clubsuit S_{ik}$

We treat the remaining pieces all at once. Figure 20.6 below shows the region of interest to us. We again call this region $\clubsuit X$. This the subset of $\clubsuit S$ lying to the right of $\clubsuit S_{14}$.

LEMMA 20.15. *$\heartsuit\Psi$ cannot intersect $\clubsuit X$.*

Proof: $\heartsuit\Psi$ cannot intersect the top edge of $\clubsuit X$, because this edge lies in the main diagonal. $\heartsuit\Psi$ cannot intersect the left edges of $\clubsuit X$ because these coincide with the right edges of $\clubsuit S_{14}$, which only intersect $\heartsuit J$ at $\mathbf{t}$. Next, $\heartsuit\Psi$ cannot intersect the right edge of $\clubsuit S$ because the only point in this edge contained in $\heartsuit J$ is a vertex of $\boldsymbol{D}$.

Now we deal with the bottom edge. The point $\mathbf{u}$ in Figure 20.6 is a symmetric cone point on $\heartsuit\partial L$. The image $\boldsymbol{H}_{111}(\mathbf{su})$ is contained in the open edge of the partition piece G that lies on the diagonal. From our proof of Theorem 1.6 we see that $\boldsymbol{H}_{111}(\mathbf{su})$ is disjoint from $\heartsuit J$. Hence the open segment $\mathbf{su}$ is disjoint from $\heartsuit J$. We have already seen that there is no point of $\heartsuit J$ lying to the right of $\mathbf{u}$. Since both $\mathbf{s}$ and $\mathbf{u}$ belong to the main component, we see that $\heartsuit\Psi$ cannot intersect the right edge of $\clubsuit X$.

Now we know that $\heartsuit\Psi \in \clubsuit X^o$. Let $h = \boldsymbol{H}_{141}$. We use the area test to check that $\det(dh) \geq \phi^{-2}$ on $\clubsuit X^o$. Hence h is a local diffeomorphism on $\clubsuit X^o$. We use the confinement test to show that $h(X^o)$ lies to the left of the y-axis and below the

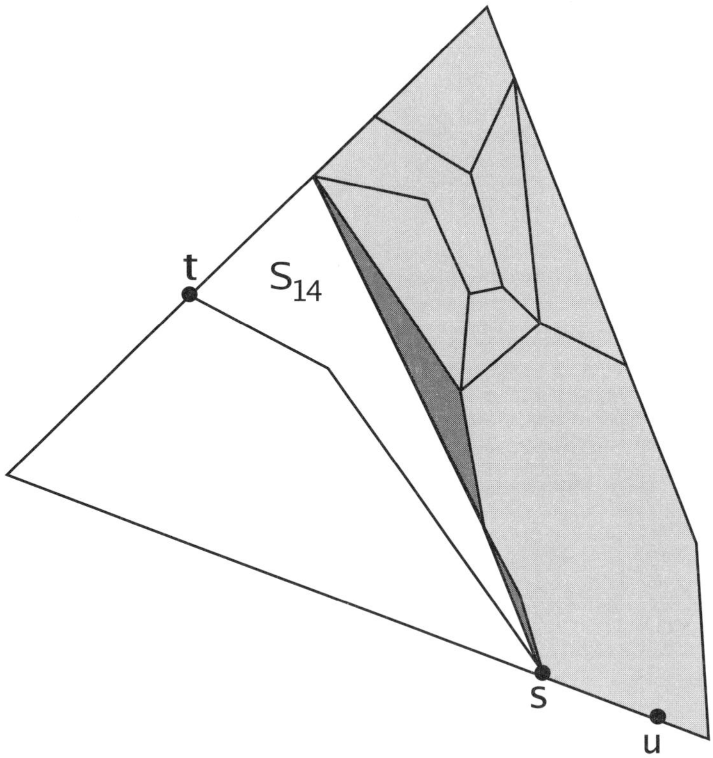

FIGURE 20.6. The region $\clubsuit X$

line λ_3 from §18.4. This region contains no blow-up points. Hence h is clean on $\clubsuit X^o$.

The C^1 loop trick finishes the proof. $\square$

We have ruled out all the possibilities, so $\heartsuit\Psi$ cannot exist. Hence $\heartsuit J$ is path connected.

20.4. The Final Picture

Figure 20.7 unites all our previous schematic pictures of $\heartsuit J$ sitting inside $\boldsymbol{M}$. The space $\boldsymbol{M}$ is the identification space formed from the 5-gon and the 15-gon using the indicated edge gluings. Compare Figure 10.5. Some subtle shading indicates the fundamental domain $\boldsymbol{T}$. $\heartsuit J$ comes in two pieces, $\heartsuit S$ and $\heartsuit J \cap L$. Here $L = \text{Fill}(\heartsuit G)$ is the filled-in $\heartsuit G$. The two pieces of $\heartsuit J$ meet at the 5 black vertices of L, the orbit $\boldsymbol{\Gamma}(1,1)$. The piece $\heartsuit S$ is the blown-down solenoid, and the piece $\heartsuit J \cap L$ is a kind of thickening of the infinite graph $\heartsuit G$ obtained, roughly speaking, by attaching additional Cantor bands to the vertices of $\heartsuit G$. The star-regular class $(0,0)$ is located at the center of L and is the central point of $\heartsuit G$.

The shaded pentagon $\boldsymbol{C}$ is the set of convex classes in $\boldsymbol{M}$. The center of $\boldsymbol{C}$ is the point (∞,∞), the regular class. The shaded region $\boldsymbol{C}'$ has the property that $\boldsymbol{C} \cup \boldsymbol{C}'$ is the connected component of $\boldsymbol{M} - \heartsuit J$ which contains the regular class. The sets $\boldsymbol{C}$ and $\boldsymbol{C}'$ are attached by the gluings of the edges a, b, c, d, e.

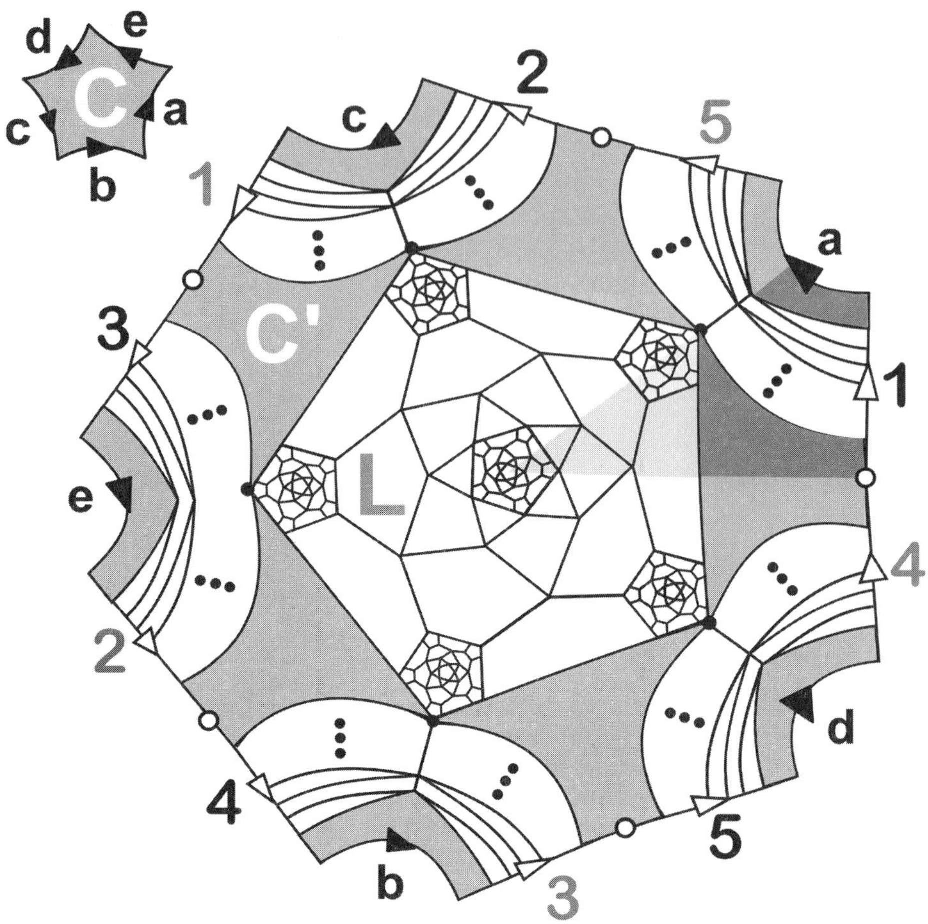

FIGURE 20.7. Schematic picture $\heartsuit J$ inside $\boldsymbol{M}$

The projective heat map carries $\partial(\boldsymbol{C} \cup \boldsymbol{C}')$ into itself. This set consists of two "pentagonal" loops which meet at their 5 vertices to form the complete graph K_5. It seems fitting that this significant subset of the Julia set makes the same pattern (without the self-intersections) as the union of a pentagon and its pentagram.

CHAPTER 21

Terms, Formulas, and Coordinate Listings

21.1. Symbols and Terms

This section functions somewhat like an index. We list (many of) the main symbols and terms used in the monograph and sometimes give a short reminder as to what they are and where they are discussed. The terms are grouped by topic rather than alphabetically.

Maps:
H: The projective heat map. See Equation 1.1.
$\boldsymbol{B}$: the change of coordinates. See Equation 1.2.
G: The Gauss recurrence. See Equation 3.16.
R: $R(x,y) = (y,x)$.
Γ: The order 10 group generated by G and R. See §3.8.
$\boldsymbol{H} = \boldsymbol{B} H \boldsymbol{B}^{-1}$.
$\boldsymbol{\Gamma} = \boldsymbol{H} \Gamma \boldsymbol{B}^{-1}$.
$\boldsymbol{G} = \boldsymbol{B} G \boldsymbol{B}^{-1}$.
$\boldsymbol{R} = \boldsymbol{B} R \boldsymbol{B}^{-1} = R$.
$\boldsymbol{H}_{ijk} = \boldsymbol{R}^i \boldsymbol{G}^j \boldsymbol{H}^k$.
Primary maps: The ten maps $\boldsymbol{H}_{ij1}$. See §10.5
S: The semigroup generated by the primary maps. See §10.5.

Sets of Polygon Classes
$\mathcal{C}$: The space of projective equivalence classes of convex pentagons. See §9
$\mathcal{C}_*$: The space of projective equivalence classes of star-convex pentagons. See §9
$\mathcal{P}$: The space of projective equivalence classes of pentagons. See §10.1.
$\mathcal{N}$: The space of projective classes of nonconvex pentagons. See §10.1
$\boldsymbol{N} = \boldsymbol{B}(\mathcal{N})$: space of nonconvex classes. See §10.2.
$\boldsymbol{T}$: Fundamental domain for the action of $\boldsymbol{\Gamma}$ acting on $\boldsymbol{N}$. See §10.2.
$\partial_0 \boldsymbol{T}$: The diagonal side of $\boldsymbol{T}$. See Figure 10.3.
$\partial_1 \boldsymbol{T}$: The vertical side of $\boldsymbol{T}$. See Figure 10.3.
$\partial_2 \boldsymbol{T}$: The remaining side of $\boldsymbol{T}$. See Figure 10.3.
M: The manifold obtained by blowing up $(\boldsymbol{R} \cup \infty)^2$ at 3 points. See §10.6.
$\boldsymbol{D}$: The big disk containing the Cantor set $\boldsymbol{B}(\mathcal{JC})$. See §12
$\boldsymbol{D}_1, \ldots, \boldsymbol{D}_6$: Smaller disk subsets of $\boldsymbol{D}$. See §12.3.
$\boldsymbol{K}_1, \ldots, \boldsymbol{K}_6$: Subsets of $\boldsymbol{D}_1, \ldots, \boldsymbol{D}_6$ which map to $\boldsymbol{D}$. See §12.5.
$\boldsymbol{M}$: The global space in the $\boldsymbol{B}$-coordinates – i.e. $\boldsymbol{B}(M)$. See §15.1.
L: Canonical subset of $\boldsymbol{D}$; filled-in version of $\heartsuit G$. See §19.
$L_1, \ldots, L_6$: Pre-images of L under $\boldsymbol{H}$. See §19.
$\boldsymbol{\Omega}$: Fundamental domain for the action of $\boldsymbol{\Gamma}$ on $\boldsymbol{D} - L$. See §20.1.

Subsets of the Julia Set
$\mathcal{J}$: The Julia set; points in $\mathcal{P}$ which do not converge to $\mathcal{C}$ under iteration. See §1.3.
$\mathcal{J}C$: The invariant Cantor set from Theorem 1.5.
$\mathcal{J}A$: The invariant Cantor band from Theorem 1.9.
$\mathcal{G}_\infty$: The infinite graph from Theorem 1.9.
$\heartsuit J$: This is the closure of $\boldsymbol{B}(\mathcal{J})$, the subject of most of our theorems.
$\heartsuit A$: This is the Cantor band derived from Theorem 14.1.
$\heartsuit A^*$: This is the Cantor band which extends $\heartsuit A$. See §16.1.
$\heartsuit JA = \boldsymbol{\Gamma}(\heartsuit A) = \boldsymbol{B}(\mathcal{J}A)$.
$\heartsuit S$: The blown-down solenoid from Theorem 1.10. See §17.
$\heartsuit W$: The canonical loop in $\heartsuit J$. See §17.4.
$\heartsuit W_1$: One fifth of $\heartsuit W$. See §19.1.
$\heartsuit X$: An arc of $\heartsuit Y$ called "the generator". See §19.1.
$\heartsuit Y$: an arc in the boundary of $\heartsuit L$. See §19.2.
$\heartsuit \partial L$: The outer pentagonal cycle of $\mathcal{G}$. See §19.3.
$\heartsuit G = \boldsymbol{B}(\mathcal{G})$, the image of the infinite graph from Theorem 1.9. See §19.8.

Some Topological Spaces:
Cantor band: Homeomorphic to the product of a Cantor set and $(0,1)$.
Cantor product: Homeomorphic to the product of a Cantor set and $[0,1]$.
Cantor cone: Homeomorphic to the cone on a Cantor set.
basic cone point: The blowdown of a Cantor product. See §15.3.
Smale Horseshoe: Classic hyperbolic dynamical system. See §6.4.
Quasi-Horseshoe: A generalization of the Smale horseshoe. See §6.5.
2-adic Solenoid. Mapping torus for the $+1$ map on the 2-attics. See §6.6.
BJK Continuum: 2-fold quotient of the solenoid. See §6.7.

Projective Geometry Notions:
polarity, duality: isomorphisms from the projective plane to its dual. See §3.3.
Cross Ratio Principle: connection between cross ratios of points and lines. See §3.4.
vertex, edge, flag invariants: projective invariants of a polygon. See §3.6.
Gauss Recurrence: A famous order 5 birational map of the plane. See §3.8.
Gauss Group: group generated by the Gauss recurrence and reflection. See §3.8.
Hilbert Metric: Projectively natural metric on a convex domain. See §3.5.
pentagram map: A projectively natural polygon iteration. See §5.
twisted polygon: A generalization of a polygon. See §5.6.

Computational Tests:
positivity criterion. See §9.3.
positive dominance algorithm. See §11.
denominator test: Tests that a function does not blow up in a domain. See §11.3.
area test: Gives lower bound on the Jacobian of a map. See §11.4.
expansion test: Gives lower bound for metric expansion of a map. See §11.5.
confinement test: Checks that $F(X) \subset Y$. See §11.6.
exclusion test: Checks that $F(X) \subset \boldsymbol{R}^2 - Y$. See §11.7.
cone test: Checks that dF maps one cone into another. See §11.8
Stretch Test: Checks expansion properties of dF. §11.9

21.2. Two Important Numbers

We have the two numbers

$$\phi = \frac{1+\sqrt{5}}{2}, \qquad \psi = \frac{1+\sqrt{13}}{2}. \tag{21.1}$$

We use ϕ all over the place and ψ occasionally. Often when we list the coordinates of our partition pieces, we use the notation

$$(a,b,c,d) = (a+b\phi, c+d\phi). \tag{21.2}$$

We will use this notation rather robotically, so that e.g. $(1,0,1,0)$ denotes the point $(1,1)$.

21.3. The Maps

Here is the projective heat map.

$$H(x,y) = (x', y'),$$

$$x' = \frac{\left(xy^2+2xy-3\right)\left(x^2y^2-6xy-x+6\right)}{\left(xy^2+4xy+x-y-5\right)\left(x^2y^2-6xy-y+6\right)}$$

$$y' = \frac{\left(x^2y+2xy-3\right)\left(x^2y^2-6xy-y+6\right)}{\left(x^2y+4xy-x+y-5\right)\left(x^2y^2-6xy-x+6\right)} \tag{21.3}$$

The Gauss group is generated by the elements

$$G(x,y) = \left(y, \frac{1-x}{1-xy}\right), \qquad R(x,y) = (y,x). \tag{21.4}$$

The conjugating map is

$$\boldsymbol{B}(x,y) = (b(x), b(y)), \qquad b(t) = \phi^3\left(\frac{\phi+t}{-1+\phi t}\right) \tag{21.5}$$

Once we have this map, we set $\boldsymbol{F} = \boldsymbol{B}F\boldsymbol{B}^{-1}$ for each map F we consider. The Gauss group $\boldsymbol{\Gamma}$ is the order 10 group generated by $\boldsymbol{G}$ and $\boldsymbol{R}$. The semigroup $\boldsymbol{S}$ is generated by the 20 elements

$$\boldsymbol{H}_{ijk} = \boldsymbol{R}^i\boldsymbol{G}^j\boldsymbol{H}^k, \qquad i \in \{0,1\}, \qquad j \in \{0,1,2,3,4\}, \qquad k \in \{0,1\}. \tag{21.6}$$

When $k=0$ these are elements of $\boldsymbol{\Gamma}$. When $k=1$ these are called *primary maps*.

21.4. Some Special Points

The vertices of the fundamental domain $\boldsymbol{T}$ are

$$\boldsymbol{p} = (\phi^6, \phi^6), \qquad (\phi^6, -\phi^{-4}), \qquad (\phi^{-4}, \phi^6). \tag{21.7}$$

The manifold $\boldsymbol{M}$ is obtained by blowing up $(\boldsymbol{R}\cup\infty)^2$ at the three points

$$\boldsymbol{p}, \qquad (-\phi^4, \phi^2), \qquad (\phi^2, -\phi^4). \tag{21.8}$$

The Julia set $\boldsymbol{J}$ has 4 special cone points

$$\boldsymbol{q} = (b(\psi), b(\psi)), \qquad \boldsymbol{r} = (\phi^2, \phi^2) = (1,1,1,1)$$

$$\mathbf{s} = (-\phi^2 b(1-\psi), b(1-\psi)), \qquad \mathbf{t} = (\phi^{-2}, \phi^{-2}) = (2,-1,2,-1) \tag{21.9}$$

21.5. The Cantor Set Pieces

Here are the partition pieces associated to the proof of Theorem 1.5.

$$\clubsuit R = (0,0,0,0)(15,-9,-39,24)(5,-3,5,-3)$$

$$\clubsuit S_1 = (\boldsymbol{s})(15,-9,-39,24)(5,-3,5,-3)$$
$$(33,-20,33,-20)(-12,8,-16,10)(-7,5,11,-7)$$

$$\clubsuit S_2 = (-5,4,14,-9)(\boldsymbol{s})(-7,5,11,-7)(-12,8,-16,10)$$
$$(27,-16,-11,7)(11,-6,-3,2)(16,-9,-5,3)$$

$$\clubsuit S_3 = (11,-6,-3,2)(14,-8,-4,3)(27,-16,-11,7)$$

$$\clubsuit T_1 = (22,-13,2,-1)(-25,16,20,-12)(33,-20,33,-20)(-12,8,-16,10)$$

$$\clubsuit T_2 = (-20,13,36,-22)(27,-16,-11,7)(14,-8,-4,3)(22,-13,33,-20)$$

$$\clubsuit T_3 = (1,0,1,0)(14,-8,-4,3)(22,-13,33,-20)(4,-2,4,-2)$$

$$\clubsuit T_4 = (22,-13,33,-20)(4,-2,4,-2)(33,-20,33,-20)$$
$$(-25,16,20,-12)(22,-13,2,-1)(-20,13,36,-22)$$

$$\clubsuit T_5 = (27,-16,-11,7)(-12,8,-16,10)(22,-13,2,-1)(-20,13,36,-22)$$

$$\spadesuit U = (1,0,1,0)(9,-7,-9,7)(2,-3,13,-7)(2,-3,-2,-3)(-20,15,-2,-3).$$

21.6. The Horseshoe Pieces

Here are the regions used in the proof of Theorem 1.6.

21.6.1. The Quasi Horseshoe. Here are the coordinates of the 8 pieces comprising the quasi-horseshoe domain.

$$\clubsuit P_1^1 = (-10,11,5,0)(17,-5,-5,6)(9,0,0,3)(-10,11,2,2)$$

$$\clubsuit P_2^1 = (-10,11,42,-23)(-1,6,-2,4)(17,-5,-5,6)(-10,11,5,0)$$

$$\clubsuit P_3^1 = (5,0,0,1)(12,-3,8,-4)(1,4,-1,2)(5,0,-1,2)$$

$$\clubsuit P_4^1 = (5,0,-7,5)(2,3,-7,5)(12,-3,8,-4)(5,0,0,1)$$

$$\clubsuit P_1^2 = (17,-5,-5,6)(6,3,14,-6)(-2,8,11,-4)(9,0,0,3)$$

$$\clubsuit P_2^2 = (-1,6,-2,4)(1,6,4,0)(6,3,14,-6)(17,-5,-5,6)$$

$$\clubsuit P_3^2 = (12,-3,8,-4)(3,4,-5,4)(5,3,-1,2)(1,4,-1,2)$$

$$\clubsuit P_4^2 = (2,3,-7,5)(6,2,-7,5)(3,4,-5,4)(12,-3,8,-4)$$

21.6.2. The Outer Layer.

$$\clubsuit A = (5,8,6,0)(5,8,5,8)(qq)$$

$$\clubsuit B \cup \clubsuit C = (3,1,-2,-3)(5,8,-2,-3)(5,8,9,-5)(6,-2,-14,8)$$

$$\clubsuit D = (-250,0,-12,8)(-2,-3,-12,8)(-4,0,122,-75)$$
$$(122,-75,-4,0)(-12,8,-2,-3)(-12,8,-250,0)(-250,0,-250,0)$$

$$\spadesuit E = (0,1,1,-1)(6,-2,-14,8)(3,1,-2,-3)(-20,15,-2,-3)$$

$$\clubsuit F = (1,1,-7,5)(5,8,-7,5)(5,8,9,-5)(6,-2,-14,8)(0,1,1,-1)(11,-6,-3,2)$$

$$\clubsuit G = (5,0,-1,2)(5,8,-1,2)(5,8,-8,7)(1,6,4,0)$$
$$(-1,6,-2,4)(-10,11,42,-23)(2,3,28,-14)(3,2,14,-5)(\boldsymbol{q})(1,1,1,1)$$

$$\clubsuit H = (5,8,6,0)(\boldsymbol{q})(58,-32,-49,34)(2,3,-38,27)$$
$$(-10,11,2,2)(9,0,0,3)(-2,8,11,-4)(5,8,1,2)$$

Our program treats $\clubsuit B$ and $\clubsuit C$ as a single piece.

21.6.3. The Inner Layer.

$$\clubsuit I_1 = (6,3,14,-6)(5,8,-14,11)(5,8,1,2)(-2,8,11,-4)$$

$$\clubsuit I_2 = (1,6,4,0)(5,8,-8,7)(5,8,-14,11)(6,3,14,-6)$$

$$\clubsuit I_3 = (3,4,-5,4)(5,8,16,-9)(5,8,-1,2)(5,3,-1,2)$$

$$\clubsuit I_4 = (6,2,-7,5)(5,8,-7,5)(5,8,16,-9)(3,4,-5,4)$$

$$\clubsuit J_1 = (2,3,-1,4)(-10,11,5,0)(-10,11,2,2)(2,3,-38,27)$$

$$\clubsuit J_2 = (2,3,28,-14)(-10,11,42,-23)(-10,11,5,0)(2,3,-1,4)$$

$$\clubsuit J_{31} = (6,-2,4,-1)(4,0,0,1)(5,0,0,1)(5,0,-1,2)(1,1,1,1)$$

$$\clubsuit J_{32} = (6,-2,4,-1)(4,0,0,1)(1,1,0,1)(1,1,1,1)$$

$$\clubsuit J_4 = (1,1,-7,5)(5,0,-7,5)(5,0,0,1)(1,1,0,1)$$

$$\clubsuit K_1 = (58,-32,27,-13)(2,3,-1,4)(2,3,-38,27)(58,-32,-49,34)$$

$$\clubsuit K_2 = (3,2,14,-5)(2,3,28,-14)(2,3,-1,4)(58,-32,27,-13)$$

$$\clubsuit L_1 = (qq)(58,-32,27,-13)(58,-32,-49,34)$$

$$\clubsuit L_2 = (qq)(3,2,14,-5)(58,-32,27,-13)$$

$$\spadesuit Y_1 = (3,1,-2,-3)(3,0,-15,8)(15,0,15,0)(5,8,5,8)$$

$$\spadesuit Y_2 = (5,8,-2,-3)(3,1,-2,-3)(5,-1,13,-10)(12,0,12,0)(5,8,5,8).$$

21.6.4. The Last Three Pieces. Here are the coordinates of the last 3 pieces of the partition.

$$\clubsuit M = (0,1,1,-1)(-5,4,14,-9)(16,-9,-5,3)$$

$$\clubsuit N = (1,0,1,0)(11,-6,-3,2)(1,1,-7,5)$$

$$\clubsuit O = (1,0,1,0)(1,1,-7,5)(1,1,1,1)$$

Here are the coordinates of the auxiliary quadrilateral used for $\clubsuit M$.

$$\spadesuit Y_3 = (16,-6,16,-6)(9,0,0,3)(5,8,1,2),(12,0,12,0).$$

21.6.5. The Two Bridges. In §14.2 we introduced two more quadrilaterals to help us with our analysis. Here they are.

$$\spadesuit Y^1 = (4,0,12,-10)(15,-7,11,-9)(12,0,12,0)(24,-7,12,0).$$

$$\spadesuit Y^2 = (3,2,-4,0)(4,0,-4,0)(15,0,15,0)(7,6,15,0).$$

21.7. The Refinement

In our proof of Theorems 1.8, 1.9, and 1.11 we needed to subdivide the piece $\clubsuit S_1$ into 7 pieces. Here they are.

$$\clubsuit S_{11} = (-12, 8, -16, 10)(33, -20, 33, -20)(145, -89, 118, -73)(-7, 5, 11, -7)$$

$$\clubsuit S_{12} = (7, -4, -16, 10)(2, -1, 2, -1)(5, -3, 5, -3)$$

$$(15, -9, -39, 24)(38, -23, -99, 61)(12, -7, 42, -26)$$

$$\clubsuit S_{13} = (15, -9, -11, 7)(2, -1, 2, -1)(12, -7, -3, 2)(119, -73, 21, -13)(\boldsymbol{s})$$

$$\clubsuit S_{14} = (33, -20, 33, -20)(145, -89, 118, -73)$$

$$(\boldsymbol{s})(119, -73, 21, -13)(12, -7, -3, 2)(2, -1, 2, -1)$$

$$\clubsuit S_{15} = (25, -15, -68, 42)(2, -1, 2, -1)(15, -9, -11, 7)(\boldsymbol{s})$$

$$\clubsuit S_{16} = (-7, 5, 11, -7)(145, -89, 118, -73)(\boldsymbol{s})$$

$$\clubsuit S_{17} = (25, -15, -68, 42)(38, -23, -99, 61)$$

$$(12, -7, 42, -26)(7, -4, -16, 10)(2, -1, 2, -1)$$

21.8. Auxiliary Polygons

Here we list the two auxiliary polygons used in our proof of Theorem 1.8 which we did not explain in the text.

$$\spadesuit Y = (-6, 5, 4, -2)(31, -18, 4, -2)(-44, 28), (29, -17)(8, -4, 29, -17).$$

$$\spadesuit Z = (87, -52, 26, -13)(33, -17, 0, 3)(-19, 19, -7, 3)(20, -7, -7, 3).$$

References

[**A1**] Z. Arai, *On Hyperbolic Plateaus of the Hénon Map*, Experimental Math. **16** (2), 2007, pp. 181–188

[**A2**] Z. Arai, *On Loops in the Hyperbolic Locus of the Complex Hénon Map and their Monodromies*, 2015, to appear in Physica D

[**AGZ**] V. I. Arnold, S. M. Gusein-Zade, A. N. Varchenko, *Singularities of Differentiable Maps, Vol 1*, Springer, 1985

[**BD**] E. Bedford, J. Diller, *Dynamics of a 2 Parameter Family of Plane Rational Maps*, J. Geom. Anal. 16.3, 2006, pp. 409–430

[**BDM**] A. Bonifant, M Dabija, J. Milnor, *Elliptic Curves as Attractors in $\boldsymbol{P}^2$, Part I: Dynamics*, Experimental Math **16.4**, 2007, pp. 385–420.

[**BLR**] P. Bleher, M. Lyubich, and R. Roeder, *Lee-Yang Zeros for DHL and 2D Rational Dynamics*, math arXiv 10094691

[**BLS**] E. Bedford, M. Lyubich, and J. Smillie, *Polynomial Diffeomorphisms of C^2: currents, equilibrium measure, and hyperbolicity*, Invent. Math. **103**, 1991, pp. 69–99

[**BS**] E. Bedford, J. Smillie, *The Henon Family: The Complex Horseshoe Locus and Real Parameter Space*, Math arXiv 0501475

[**Bo**] A. Bogomolny, *Cut The Knot*, http://www.cut-the-knot/ctl/Napolegon.shtml

[**F**] W. Fulton, *Algebraic Curves: An Introduction to Algebraic Geometry*, online version: http://www.math.lsa.umich.edu/~wfulton/CurveBook.pdf, 2008

[**Gli**] M. Glick, *The pentagram map and Y-patterns*, Adv. Math. **227**, 2012, pp. 1019–1045.

[**GP**] V. Guillemin and A. Pollack, *Differential Topology*, Prentice-Hall 1974.

[**Gr**] Br. Grünbaum, *Metamorphoses of polygons*, in The Lighter Side of Mathematics, Proc. Eugène Strens Memorial Conference, edited by R. K. Guy and R. E. Woodrow, M.A.A. 1994, pp. 35–48

[**GSTV**] M. Gekhtman, M. Shapiro, S.Tabachnikov, A. Vainshtein, *Higher pentagram maps, weighted directed networks, and cluster dynamics*, Electron. Res. Announc. Math. Sci. **19**, 2012, pp. 1–17

[**HK**] B. Hasselblatt and A. Katok, *Introduction to the Modern Theory of Dynamical Systems*, Cambridge University Press, 1997

[**HP**], J. Hubbard and P. Papadopol, *Newton's method applied to two quadratic equations in $\boldsymbol{C}^2$ viewed as a global dynamical system*, Memoirs of the A.M.S., 191/891, 2008

[**K**] K. Kendig, *Elementary Algebraic Geometry*, Graduate Texts in Mathematics, Springer 1977.

[**KS**] B. Khesin, F. Soloviev *Integrability of higher pentagram maps*, Mathem. Annalen., vol. 357, 2013, pp. 1005–1047

[**MB1**] G. Mari Beffa, *On Generalizations of the Pentagram Map: Discretizations of AGD Flows*, Journal of Nonlinear Sciance vol. 23, issue 2, 2013, pp. 303–334

[**MB2**] G. Mari Beffa, *On integrable generalizations of the pentagram map*, Int. Math. Res. Notices (12), 2015, pp. 3669–3693

[**Mot**] Th. Motzkin, *The pentagon in the projective plane, with a comment on Napier's rule*, Bull. Amer. Math. Soc. **52**, 1945, pp. 985–989.

[**N**] V. Nekrashevych, *The Julia set of a post-critical endomorphism of $\boldsymbol{CP}^2$*, J. Modern Dyn., 6.3, 2015 pp. 327–375

[**OST1**] V. Ovsienko, R. Schwartz, S. Tabachnikov, *The pentagram map: A discrete integrable system*, Comm. Math. Phys. **299**, 2010, pp. 409–446.

[**OST2**] V. Ovsienko, R. Schwartz, S. Tabachnikov, *Liouville-Arnold integrability of the pentagram map on closed polygons*, Duke Math. Journal vol. 162 no. 15, 2013, pp. 1149–1196

[**S1**] R. E. Schwartz, *The pentagram map*, Experimental Math. **1**, 1992, pp. 71–81

[**S2**] R. E. Schwartz, *The pentagram map is recurrent*, Experimental Math. **10**, 2001, pp. 519–528

[**S3**] R. E. Schwartz, *Discrete monodromy, pentagrams, and the method of condensation*, J. of Fixed Point Theory and App.l. **3**, 2008, pp. 379–409

[**S4**] R. E. Schwartz, *A Conformal Averaging Process on the Circle* Geom. Dedicata., 117.1, 2006.

[**Sol**] F. Soloviev *Integrability of the Pentagram Map*, Duke Math Journal vol. 162, no. 15,, 2013 pp. 2815–2853

[**ST**] R. Schwartz and S. Tabachnikov, *Elementary surprises in projective geometry*, Math. Intelligencer 32(3), 2010, pp. 31–34

[**Tu**] W. Tucker, *The Lorenz attractor exists*, C. R. Acad. Sci. Paris t. 328, Série 1, 1999, pp. 1197–1202

[**W**] *Mathematica*, Wolfram Research, www.wolfram.com

Selected Published Titles in This Series

219 **Richard Evan Schwartz,** The Projective Heat Map, 2017

215 **Robert Bieri and Ralph Strebel,** On Groups of PL-homeomorphisms of the Real Line, 2016

214 **Jared Speck,** Shock Formation in Small-Data Solutions to 3D Quasilinear Wave Equations, 2016

213 **Harold G. Diamond and Wen-Bin Zhang (Cheung Man Ping),** Beurling Generalized Numbers, 2016

212 **Pandelis Dodos and Vassilis Kanellopoulos,** Ramsey Theory for Product Spaces, 2016

211 **Charlotte Hardouin, Jacques Sauloy, and Michael F. Singer,** Galois Theories of Linear Difference Equations: An Introduction, 2016

210 **Jason P. Bell, Dragos Ghioca, and Thomas J. Tucker,** The Dynamical Mordell–Lang Conjecture, 2016

209 **Steve Y. Oudot,** Persistence Theory: From Quiver Representations to Data Analysis, 2015

208 **Peter S. Ozsváth, András I. Stipsicz, and Zoltán Szabó,** Grid Homology for Knots and Links, 2015

207 **Vladimir I. Bogachev, Nicolai V. Krylov, Michael Röckner, and Stanislav V. Shaposhnikov,** Fokker–Planck–Kolmogorov Equations, 2015

206 **Bennett Chow, Sun-Chin Chu, David Glickenstein, Christine Guenther, James Isenberg, Tom Ivey, Dan Knopf, Peng Lu, Feng Luo, and Lei Ni,** The Ricci Flow: Techniques and Applications: Part IV: Long-Time Solutions and Related Topics, 2015

205 **Pavel Etingof, Shlomo Gelaki, Dmitri Nikshych, and Victor Ostrik,** Tensor Categories, 2015

204 **Victor M. Buchstaber and Taras E. Panov,** Toric Topology, 2015

203 **Donald Yau and Mark W. Johnson,** A Foundation for PROPs, Algebras, and Modules, 2015

202 **Shiri Artstein-Avidan, Apostolos Giannopoulos, and Vitali D. Milman,** Asymptotic Geometric Analysis, Part I, 2015

201 **Christopher L. Douglas, John Francis, André G. Henriques, and Michael A. Hill, Editors,** Topological Modular Forms, 2014

200 **Nikolai Nadirashvili, Vladimir Tkachev, and Serge Vlăduţ,** Nonlinear Elliptic Equations and Nonassociative Algebras, 2014

199 **Dmitry S. Kaliuzhnyi-Verbovetskyi and Victor Vinnikov,** Foundations of Free Noncommutative Function Theory, 2014

198 **Jörg Jahnel,** Brauer Groups, Tamagawa Measures, and Rational Points on Algebraic Varieties, 2014

197 **Richard Evan Schwartz,** The Octagonal PETs, 2014

196 **Silouanos Brazitikos, Apostolos Giannopoulos, Petros Valettas, and Beatrice-Helen Vritsiou,** Geometry of Isotropic Convex Bodies, 2014

195 **Ching-Li Chai, Brian Conrad, and Frans Oort,** Complex Multiplication and Lifting Problems, 2014

194 **Samuel Herrmann, Peter Imkeller, Ilya Pavlyukevich, and Dierk Peithmann,** Stochastic Resonance, 2014

193 **Robert Rumely,** Capacity Theory with Local Rationality, 2013

192 **Messoud Efendiev,** Attractors for Degenerate Parabolic Type Equations, 2013

191 **Grégory Berhuy and Frédérique Oggier,** An Introduction to Central Simple Algebras and Their Applications to Wireless Communication, 2013

For a complete list of titles in this series, visit the AMS Bookstore at **www.ams.org/bookstore/survseries/**.